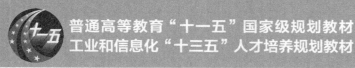

普通高等教育"十一五"国家级规划教材
工业和信息化"十三五"人才培养规划教材

计算机
专业英语 第3版

Computer Professional English

卜艳萍 周伟 ◎ 编著

人民邮电出版社
北京

图书在版编目（CIP）数据

计算机专业英语 / 卜艳萍，周伟编著. -- 3版. -- 北京：人民邮电出版社，2017.3（2021.12重印）
工业和信息化"十三五"人才培养规划教材
ISBN 978-7-115-43143-1

Ⅰ. ①计… Ⅱ. ①卜… ②周… Ⅲ. ①电子计算机－英语－高等学校－教材 Ⅳ. ①TP3

中国版本图书馆CIP数据核字(2016)第218262号

内 容 提 要

全书按照专业英语的语法知识，计算机的硬件、软件、数据库、网络、电子商务、计算机应用等内容分成 7 章。第 1 章计算机专业英语基础知识，内容包括计算机专业英语的构词法分析、语法知识介绍、专业资料的翻译与写作等内容；第 2 章计算机硬件知识，内容有中央处理器、存储器和输入/输出设备；第 3 章计算机软件知识，内容有操作系统、数据结构、编程语言；第 4 章数据库技术，内容包括数据库原理、数据仓库与数据挖掘、大数据与云计算；第 5 章计算机网络技术，内容有网络基础、信息安全、无线网络；第 6 章电子商务，内容包括电子商务、网站导航、电子支付与物流；第 7 章计算机应用，内容有办公自动化、远程教育和人工智能。每章的最后部分是专业英语应用模块。

本书可作为高等院校计算机相关专业学生的教材，也可作为计算机相关工程技术人员学习专业英语的参考资料。

◆ 编　著　卜艳萍　周 伟
　　责任编辑　范博涛
　　责任印制　焦志炜

◆ 人民邮电出版社出版发行　北京市丰台区成寿寺路 11 号
　　邮编 100164　电子邮件 315@ptpress.com.cn
　　网址 http://www.ptpress.com.cn
　　固安县铭成印刷有限公司印刷

◆ 开本：787×1092　1/16
　　印张：18.5　　　　　　　　2017 年 3 月第 3 版
　　字数：466 千字　　　　　　2021 年 12 月河北第 15 次印刷

定价：46.00 元

读者服务热线：(010)81055256　印装质量热线：(010)81055316
反盗版热线：(010)81055315
广告经营许可证：京东市监广登字20170147号

前 言

计算机专业英语是计算机相关专业的专业知识综合课程。学生在专业课学习的基础上，通过本课程的学习，能够掌握必要的计算机专业词汇和专业文献的阅读和理解能力。

在计算机专业英语（第2版）教材的基础上，调整了部分课文及阅读材料的范围和内容，增加了大数据、云计算、移动商务、物联网、网络社区、数据仓库、数据挖掘、电子支付、物流、3D打印技术等新技术的内容。对全书的结构也做了些调整，每章内容更专注于某一计算机领域的知识，便于知识点的掌握和分解。每章的最后增加了专业英语应用模块，内容包括BIOS功能、密码设置、计算机故障分析，以及如何编写简历、自荐材料、咨询信、通知信、预约便条等文档。

本书的编者均是长期从事计算机及相关专业课程教学的教师，在编写过程中能注重选择专业核心知识。全书写作风格简明扼要，理论与实践相结合。课文与阅读材料的选材上紧跟计算机专业的发展特点以及实际应用情况，既考虑与计算机专业基础课、专业课的结合，又尽可能囊括计算机科学的前沿技术和知识。每篇课文都有单词解释、难点注释和三种题型的练习，书末附有参考译文、练习答案和专业技术词汇表。

本书由上海交通大学的卜艳萍副教授和华东理工大学的周伟副教授共同编写。卜艳萍编写了第1章至第3章、附录1、附录2，并负责全书的统稿工作。周伟编写了第4章至第7章内容。在本书编写过程中，得到赵桂钦、王德俊、何飞、卜艳慧等人的指导和建议，邱遥、周烨晴同学帮助编辑及整理了部分书稿。在此对以上各位表示衷心的感谢。

由于作者学识有限，书中难免有不当之处。竭诚希望各位同行、专家及广大读者指正。

作　者
2016年12月于上海

目录 CONTENTS

第1章 计算机专业英语基础知识 1

- 1.1 计算机专业英语的特点 1
  - 1.1.1 用词和语法的特点 1
  - 1.1.2 It … 句型结构 3
  - 1.1.3 专业文献的特征 3
- 1.2 计算机专业词汇的构词法分析 5
  - 1.2.1 专业英语词汇的构成特点 5
  - 1.2.2 词汇缩略 9
  - 1.2.3 计算机专用术语与命令 10
- 1.3 专业英语中的常用语法知识 13
  - 1.3.1 动词不定式 13
  - 1.3.2 分词 16
  - 1.3.3 动名词 17
  - 1.3.4 被动语态 18
  - 1.3.5 定语从句 20
  - 1.3.6 状语从句 22
  - 1.3.7 时态简介 24
- 1.4 计算机专业资料的翻译与写作 27
  - 1.4.1 专业英语的阅读 27
  - 1.4.2 专业英语翻译 29
  - 1.4.3 专业英语写作 37

第2章 Hardware Knowledge 39

- 2.1 CPU 39
  - 2.1.1 Text 39
  - 2.1.2 Exercises 44
  - 2.1.3 Reading Material Computer Hardware Basics 45
- 2.2 Memory 47
  - 2.2.1 Text 47
  - 2.2.2 Exercises 51
  - 2.2.3 Reading Material Accessing Memory 52
- 2.3 Input/Output Devices 53
  - 2.3.1 Text 53
  - 2.3.2 Exercises 58
  - 2.3.3 Reading Material Building a Computer 60
- 2.4 专业英语应用模块 62
  - 2.4.1 The Function of BIOS 62
  - 2.4.2 Password Setup 62

第3章 Software Knowledge 64

- 3.1 Operating System 64
  - 3.1.1 Text 64
  - 3.1.2 Exercises 68
  - 3.1.3 Reading Material Linux Operating System 69
- 3.2 Data Structures 71
  - 3.2.1 Text 71
  - 3.2.2 Exercises 74
  - 3.2.3 Reading Material Stacks and Queues 76
- 3.3 Programming Language 77
  - 3.3.1 Text 77
  - 3.3.2 Exercises 80
  - 3.3.3 Reading Material Software Engineering 82
- 3.4 专业英语应用模块 83
  - 3.4.1 Computer Malfunction 83
  - 3.4.2 The Malfunctions of Main Board 84

第4章 Database Technology 86

- 4.1 Database Principle 86
  - 4.1.1 Text 86
  - 4.1.2 Exercises 90
  - 4.1.3 Reading Material Introduction to Typical Databases 91
- 4.2 Data Warehouse and Data Mining 93
  - 4.2.1 Text 93
  - 4.2.2 Exercises 97
  - 4.2.3 Reading Material Expert System 99
- 4.3 Big Data and Cloud Computing 100

	4.3.1	Text	100	4.4	专业英语应用模块	107
	4.3.2	Exercises	104		4.4.1 The Interview Questions	107
	4.3.3	Reading Material Top Technologies of Computer Science	106		4.4.2 Resume	108

第 5 章 Computer Network Technology 110

5.1	Computer Network Basics		110	5.3	Wireless Networks	124
	5.1.1	Text	110		5.3.1 Text	124
	5.1.2	Exercises	114		5.3.2 Exercises	128
	5.1.3	Reading Material Optical Communication	116		5.3.3 Reading Material Internet Applications	129
5.2	Information Security		117	5.4	专业英语应用模块	131
	5.2.1	Text	117		5.4.1 Self-Introduction	131
	5.2.2	Exercises	121		5.4.2 Self-Recommendation	132
	5.2.3	Reading Material Techniques for Internet Security	122			

第 6 章 Electronic Commerce 133

6.1	Electronic Commerce		133		6.3.2 Exercises	151
	6.1.1	Text	133		6.3.3 Reading Material The Internet of Things	152
	6.1.2	Exercises	138			
	6.1.3	Reading Material Business Networking Technology	139	6.4	Electronic Payment and Logistics	154
					6.4.1 Text	154
6.2	Web Navigation		141		6.4.2 Exercises	158
	6.2.1	Text	141		6.4.3 Reading Material Electronic Marketing	159
	6.2.2	Exercises	144	6.5	专业英语应用模块	161
	6.2.3	Reading Material Social Networking	145		6.5.1 Appointment Letter	161
6.3	Mobile Commerce		147		6.5.2 Letter of Inquiry	161
	6.3.1	Text	147			

第 7 章 Computer Applications 163

7.1	Office Automation		163	7.3	Artificial Intelligence	177
	7.1.1	Text	163		7.3.1 Text	177
	7.1.2	Exercises	167		7.3.2 Exercises	181
	7.1.3	Reading Material 3D Printing	169		7.3.3 Reading Material Intelligent Technologies	182
7.2	Distance Education		170	7.4	专业英语应用模块	184
	7.2.1	Text	170		7.4.1 Notification Letter	184
	7.2.2	Exercises	174		7.4.2 Letters of Apologies	185
	7.2.3	Reading Material Multimedia Technology	176			

参考译文　186

第2章	硬件知识	186
2.1	中央处理器	186
2.2	存储器	190
2.3	输入/输出设备	193
第3章	软件知识	196
3.1	操作系统	196
3.2	数据结构	199
3.3	编程语言	202
第4章	数据库技术	205
4.1	数据库原理	205
4.2	数据仓库和数据挖掘	208
4.3	大数据和云计算	212
第5章	计算机网络技术	215
5.1	计算机网络基础	215
5.2	信息安全	218
5.3	无线网络	222
第6章	电子商务	225
6.1	电子商务	225
6.2	网站导航	228
6.3	移动商务	231
6.4	电子支付与物流	234
第7章	计算机应用	236
7.1	办公自动化	236
7.2	远程教育	240
7.3	人工智能	243

练习答案　247

第2章	Hardware Knowledge	247
2.1	CPU	247
2.2	Memory	247
2.3	Input/Output Devices	248
第3章	Software Knowledge	248
3.1	Operating System	248
3.2	Data Structures	249
3.3	Programming Language	249
第4章	Database Technology	250
4.1	Database Principle	250
4.2	Data Warehouse and Data Mining	250
4.3	Big Data and Cloud Computing	250
第5章	Computer Network Technology	251
5.1	Computer Network Basics	251
5.2	Information Security	251
5.3	Wireless Networks	252
第6章	Electronic Commerce	252
6.1	Electronic Commerce	252
6.2	Web Navigation	253
6.3	Mobile Commerce	253
6.4	Electronic Payment and Logistics	254
第7章	Computer Applications	254
7.1	Office Automation	254
7.2	Distance Education	254
7.3	Artificial Intelligence	255

附录1　计算机专业英语词汇表　256

附录2　计算机专业英语缩写词表　276

参考文献　290

第 1 章 计算机专业英语基础知识

1.1 计算机专业英语的特点

科学技术本身的性质决定了专业英语与普通英语（Common English）有很大的差异。科技专业英语是在自然科学和工程技术领域使用的一种英语文体，是随着科学技术的迅速发展而逐渐形成的。在词汇上，科技专业英语中含有大量的专业技术词汇和术语；在语法应用上，科技专业英语中常用被动语态、非谓语动词、名词化结构和从句等。科技专业英语文体的特点是：清晰、精练、严密、准确。

计算机行业是一个充满活力和不断创新的领域。软、硬件技术方面的新思想和新概念层出不穷，最新的计算机专业软件及图书通常是用英语书写和描述的，正确理解和翻译这些新的专业知识和术语是非常重要的。

我们经常接触到计算机教科书、参考书、说明书，人机交互式的键盘键名，屏幕上的窗口菜单、浏览器上的各种网页，各种操作系统的命令、各种程序语言中的语句，这些都需要我们掌握一定的计算机专业英语知识。而要学好计算机专业英语知识，就必须掌握其特点。计算机专业是融科学性与技术性于一体的学科，其专业文献必须遵循科技文体的规范，因而计算机专业英语在用词与语法等方面具有一些显著的特点。

1.1.1 用词和语法的特点

专业英语的主要特点是它具有很强的专业性，这主要体现在它的特殊专业内容和特殊专业词汇上。词汇是组成句子的基本元素，对词汇含义不能确定，就很难理解句子的内容。

计算机专业英语文献中专业术语多，而且派生和新出现的专业用语还在不断地增加。另外，计算机专业英语文献中的缩略词汇多，而且新的缩略词汇还在不断增加，并成为构成新词的词源。如 CPU（Central Processing Unit）、WPS（Word Processing System）、NT（Net Technology）、IT（Information Technology）等，掌握这些词汇首先要有一定的英语词汇量，还要对新技术有所了解。

概括起来，计算机专业英语在用词中具有以下一些特点。

- 名词性词组多；

- 合成新词多；
- 介词短语多；
- 非限定动词（尤其是分词）使用频率高；
- 半技术词汇多；
- 缩略语使用频繁；
- 单个动词比动词词组用得频繁；
- 常使用动词或名词演化成的形容词；
- 希腊词根和拉丁词根比例大。

例：This approach mitigates complexity separating the concerns of the front end, which typically revolve around language semantics, error checking, and the like, from the concerns of the back end, which concentrates on producing output that is both efficient and correct.

译文：这种方法减轻了把处理语义、检测错误等前端工作和主要产生正确有效输出的后端工作分离开来的复杂性。

例：Alternative calculation models in neural networks include models with loops, where some kind of time delay process must be used, and "winner takes all" models, where the neuron with the highest value from the calculation fires and takes a value 1, and all other neurons take the value 0.

译文：神经网络的另外计算模型包括带有回路的模型和"胜者通吃"模型，其中带回路模型必须使用某种时间延迟处理，而在"胜者通吃"模型中，从计算中得到最大值的神经元触发并赋值 1，所有其他的神经元赋值 0。

上面这两句话描述的是计算机专业领域内的知识，如果不了解编译原理和神经网络技术及其相关词汇，则很难给出准确的翻译。

由于科学技术关心的不是个人的心理情绪，而是客观的普遍规律和对过程、概念的描述，因此专业英语应具有的客观性和无人称性必然要反映到语法结构上来。

专业英语往往在句子结构上采用被动语态描述，即以被描述者为主体，或者以第三者的身份介绍文章要点和内容。于是，被动语态反映了专业英语文体的客观性。除了表述作者自己的看法、观点以外，很少直接采用第一人称表述法，但在阅读理解和翻译时，根据具体情况，又经常要将一个被动语态句子翻译成主动形式，以便强调某个重点，同时更适合汉语的习惯。

计算机专业英语在语法上有如下显著特点。

- 专业术语多；
- 常用 It ... 句型结构；
- 长句多；
- 被动语态使用频繁；
- 在说明书、手册中广泛使用祈使语句；
- 用虚拟语气表达假设或建议。

例：Were there no plants, there would be no photosynthesis and life could not go on.

译文：如果没有植物，就没有光合作用，生命就无法继续下去。

在说明事理并涉及到各种前提和条件时，可以用虚拟语气。

例：Backing up your files safeguards them against loss if your hard disk fails or you accidentally overwrite or delete data.

译文：当硬盘发生故障或用户意外覆盖、删除数据时，备份可以保护文件，避免损失。

例：Must be structure field name.

译文：需要的是结构字段名。

祈使语句常用来表示指示、建议、劝告和命令等意思，可以用于说明书、操作规程和注意事项等资料中。

1.1.2 It ... 句型结构

It ... 结构在专业英语中用的较多，下面列出常用的句型。

- It is +名词+从句

It is a fact that ...　事实是……

It is a question that ...　……是个问题

It is no wonder that ...　毫无疑问……

It is the law of nature that ...　……是自然规律

It is a common practice that ...　通常的做法是……

- It is +形容词+从句

It is necessary that ...　有必要……

It is clear that ...　很清楚……

It is important that ...　重要的是……

It is natural that ...　很自然的是……

- It is +过去分词+从句

It is said that ...　据说……

It is believed that ...　确信……

It has been proved that ...　已证明……

It is generally considered that ...　人们普遍认为……

- It is +介词短语+从句

It is from this point of view that ...由此看来……

It is of great significance ...　……具有重大的意义

It is only under these conditions that ...　只有在这些条件下才能……

- It +不及物动词+从句

It follows that ...　由此可见……

It turned out that ...　结果是……

It may be that ...　可能……

It stands to reason that ...　显然……

例：It was in the 1940's that the first computer was built.

译文：第一台计算机建成于20世纪40年代。

例：It is necessary to learn Visual Basic.

译文：学习 Visual Basic 是很有必要的。

1.1.3 专业文献的特征

由于各个领域的专业英语都以表达科技概念、理论和事实为主要目的，因此专业英语很注重客观事实和真理，并且要求逻辑性强，条理规范，表达准确、精练、正式。因而专业英语文献在内容上具有客观（objectivity）、精练（conciseness）和准确（accuracy）三个特征。

1．客观

因为要求客观，所以常用被动语态和一般现在时。有人统计专业英语中被动语态的句子要占 1/3～1/2。即使用主动语态，主语也常常是非动物的（inanimate subject）。

例：The procedure by which a computer is told how to work is called programming.

句子的主要结构为 The procedure is called programming。用一般现在时和被动语态。by which 为"介词+关系代词"引导定语从句，从句的谓语也为被动语态，which 指代 procedure。

译文：告诉计算机如何工作的过程称为程序设计。

就时态而言，因为专业科技文献所涉及的内容（如科学定义、定理、方程式或公式、图表等）一般并没有特定的时间关系，所以在专业文献中大部分都使用一般现在时。至于一般过去时、一般完成时也在专业英语中经常出现，如科技报告、科技新闻、科技史料等。

2．精练

因为要求精练，专业英语中常希望用尽可能少的单词来清晰地表达原意。这就导致了非限定动词、名词化单词或词组及其他简化形式的广泛使用。

通常的表达形式为：

（1）What does a fuse do? It protects a circuit.

（2）It is necessary to examine whether the new design is efficient.

精练的表达形式为：

（1）The function of a fuse is to protect a circuit.

（2）It is necessary to examine the efficiency of the new design.

3．准确

专业英语的准确性主要表现在用词上，在语法结构上也有其特点。例如，为了准确精细地描述事物过程，所用句子都较长，有些甚至一段就是一个句子。长句反映了客观事物中复杂的关系，它与前述精练的要求并不矛盾，句子长结构仍是精练的，只是包含的信息量大，准确性较高。

在掌握词汇、语法特点与文献具有的三个特征基础上，面对专业英语文献的长句首先要进行语法分析。语法分析主要从两点入手，首先是找出谓语和主语。其次是找出连接词。找出了连接词就找到了句子间的界限和它们之间的关系。这里说的连接词是指包括连接代词、连接副词、关系代词和关系副词等的广义连接词。

在英语科技文献中还常常使用多重复句，形成多个层次，以便能严谨地表达复杂的思想。如果把一句话分成几个独立的句子，就有可能影响到句子之间的密切联系。文章的论述性越强，多重复句用得越多，句子也越长。多重复句的分句之间有两种关系，一种是并列关系，另一种是主从关系，但是以主从关系为主。这两种关系常常同时出现在一个句子中。从句在说明主句的时候，有三种可能的位置：在主句前、在主句后和插在主句中间。

例：This instrument works on the principle that each individual substance emits a characteristic spectrum of light when its molecules are caused to vibrate by the application of heat, electricity, etc.; and after studying the spectrum which he had obtained on this occasion, Hildebrand reported the gas to be nitrogen.

这个句子的基本骨架是用 and 连接的两个并列的主句 This instrument works on the

principle 和 after studying the spectrum, Hildebrand reported the gas to be nitrogen。第一个主句有一个用 that 连接的同位语从句,说明主句中的 principle,而这个同位语从句又有一个用 when 连接的时间状语从句。第二个主句中包含了一个用 which 连接的定语从句,说明 spectrum。

译文:这个仪器工作的原理是,当物质的分子由于加热、通电等而引起振动时,每种物质产生一种独特的光谱;在研究了此种情况下收集到的光谱后,Hildebrand 宣布这种气体是氮。

1.2 计算机专业词汇的构词法分析

1.2.1 专业英语词汇的构成特点

词汇是语言发展的产物。语言在发展过程中,旧的词不断被抛弃,新的词不断在产生。随着科学技术的发展,新术语、新概念、新理论和新产品不断出现。不但新词(及词组)大量涌现,许多日常用语也不断增加新的科技含义,如 off-the-shelf(成品的)、state-of-the-art(现代化的)等。在专业英语中,缩略词的增加尤其迅速,各类技术词汇也随着专业的细分、学科的渗透而日益增多。在阅读英文科技文章时,我们可能会遇到许多不认识的词汇。若从字典查找费时或根本查不到合适的意义,可以通过上下文来领悟生词。

1. 专业英语中常见的词汇类型

(1)技术词汇(technical words)

这类词的意义狭窄、单一,一般只使用在各自的专业范围内,因而专业性很强。这类词一般较长并且越长词义越狭窄,出现的频率也不高。

例:bandwidth(带宽),flip-flop(触发器),superconductivity(超导性),hexadecimal(十六进制),amplifier(放大器)等。

(2)次技术词汇(sub-technical words)

次技术词汇是指词义不受上下文限制,各专业中出现频率都很高的词。这类词往往在不同的专业中具有不同的含义。

例:register 在计算机系统中表示寄存器,在电学中表示计数器、记录器,在乐器中表示音区,而在日常生活中则表示登记簿、名册、挂号信等。

(3)特用词(big words)

在日常英语中,为使语言生动活泼,常使用一些短小的词或词组。而在专业英语中,表达同样的意义时,为了准确、正式、严谨,不引起歧义却往往选用一些较长的特用词。这些词在非专业英语中极少使用但却属于非专业英语。

为了说明灯点亮了,在日常英语中常用:

Then the light is turned on.

而在专业英语中,常表述为连接灯的电路接通了,即:

The circuit is then completed.

这是由于 complete 词义单一准确,可以避免歧义。而 turned on 不仅表示开通,而且还可以表示其他意义,如:

The success of a picnic usually turns on(依赖)the weather.

类似的词还有:

go down —— depress   turn upside down —— invert

keep —— maintain   enough —— sufficient
push in —— insert   find out —— determine

（4）功能词（function words）

功能词包括介词、连词、冠词、代词等。功能词为词在句子中的结构关系提供了十分重要的结构信号，对于理解专业内容十分重要，而且出现频率极高。研究表明，在专业英语中，出现频率最高的 10 个词都是功能词，其顺序为：the，of，in，and，to，is，that，for，are，be。下例中 14 个词中功能词就占了 6 个。

When the recorder is operated in the record mode, previous recordings are automatically erased.

译文：当录音机工作在录音模式时，以前的录音被自动擦除。

2．专业英语中的词汇来源

专业英语中的词汇来源有以下几种情况。

（1）来源于英语中的普通词，但被赋予了新的词义

例：Work is the transfer of energy expressed as the product of a force and the distance through which its point of application moves in the direction of the force.

本句中的"work、energy、product、force"都是从普通词汇中借来的物理学术语。"work"的意思不是"工作"，而是"功"；"energy"的意思不是"活力"，而是"能"；"product"的意思不是"产品"，而是"乘积"；"force"的意思不是"力量"，而是"力"。

译文：功是能的传递，表达为力与力的作用点沿着力的方向移动的距离的乘积。

（2）来源于希腊语或拉丁语

例：

thermo	热（希腊语）	thesis	论文（希腊语）
parameter	参数（拉丁语）	radius	半径（拉丁语）
formula	公式（拉丁语）	data	数据（拉丁语）

（3）由两个或两个以上的单词组成合成词

合成词是专业英语中另一大类词汇，其组成面广，多数以连字符"-"连接单词构成，或者采用短语构成。合成方法有名词+名词、形容词+名词、动词+副词、名词+动词、介词+名词、形容词+动词等。但是合成词并非可以随意构造，否则会形成一种非正常的英语句子结构。虽然可由多个单词构成合成词，但这种合成方式太冗长，应尽量避免。

下面这些是由连字符"-"连接的合成词。

file + based → file-based   基于文件的
Windows + based → Windows-based   以 Windows 为基础的
object + oriented → object-oriented   面向对象的
thread + oriented → thread-oriented   面向线程的
point + to + point → point-to-point   点到点
plug + and + play → plug-and-play   即插即用
pear + to + pear → pear-to-pear   对等的
front + user → front-user   前端用户
push + up → push-up   上拉
pull + down → pull-down   下拉

paper + free → paper-free 无纸的
jumper + free → jumper-free 无跳线的
user + centric → user-centric 以用户为中心的
power + plant → power-plant 发电站
conveyer + belt → conveyer-belt 传送带

随着词汇的专用化，合成词中间的连接符被省略掉，形成了一个独立的单词。如：

in + put → input 输入
out + put → output 输出
feed + back → feedback 反馈
fan + in → fanin 扇入
fan + out → fanout 扇出
on + line → online 在线
metal + work → metalwork 金属制品

英语中有很多专业术语由两个或更多的词组成，叫作复合术语。它们的构成成分虽然看起来是独立的，但实际上合起来构成一个完整的概念。

liquid crystal   液晶
computer language   计算机语言
machine building   机器制造
linear measure   长度单位
civil engineering   土木工程

（4）派生词（derivation）

派生也叫缀合。由派生而来的专业词汇非常多，专业英语词汇大部分都是用派生法构成的，它是根据已有的词，通过对词根加上各种前缀和后缀构成的新词。这些词缀有名词词缀，如 inter-、sub-、in-、tele-、micro-等；形容词词缀，如 im-、un-、-able、-al、-ing、-ed 等；动词词缀，如 re-、under-、de-、-en、con-等。其中，采用前缀构成的单词在计算机专业英语中占了很大比例。下面是一些典型的派生词。

加前缀构成新词只改变词义，不改变词性。如：

multimedia   多媒体   multiprocessor   多处理器
interface   接口   microprocessor   微处理器
microcode   微代码   hypertext   超文本
hypermedia   超媒体   telephone   电话
teleconference 远程会议   telegraph   电报
barometer   气压表   barograph   气压记录仪
ultrasonic   超声的   subsystem   分系统
hydro-electric 发电   non-metal   非金属

英语的前缀是有固定意义的，记住其中的一些常用前缀对于记忆生词和猜测词义很有帮助。下面是一些具有否定意义的前缀。

anti-   表示"反对"   antibody   抗体
counter-   表示"反对，相反"   counterbalance   反平衡
contra-   表示"反对，相反"   contradiction   矛盾

de-	表示"减少,降低,否定"	decrease 减少
		decompose 分解
dis-	表示"否定,除去"	discharge 放电
		disassemble 拆卸
in- il-	(在字母 l 前)表示"不"	inaccurate 不准确的
		illegal 违法的
im-	(在字母 m、b、p 前)表示"不"	imbalance 不平衡的
		impure 不纯的
mis-	表示"错误"	mislead 误导
non-	表示"不,非"	non-ferrous 有色金属的
un-	表示"不、未、丧失"	unaccountable 说明不了的
		unknown 未知的

加后缀构成新词可能改变也可能不改变词义,但一般改变词性。有的派生词加后缀的时候,语音或拼写可能发生变化。从一个词的后缀可以判别它的词类,这是它的语法意义。它们的词汇意义往往并不明显。

electric	(形容词) + ity → electricity (名词:电,电学)
liquid	(名词) + ize → liquidize (动词:液化)
conduct	(动词) + or → conductor (名词:导体)
invent	(动词) + ion → invention (名词:发明)
propel	(动词) + l + er → propeller (名词:推进器)
simple	(形容词) + icity → simplicity (名词:单纯,注意拼写有变化)
maintain	(动词) + ance → maintenance (名词:维修,注意拼写有变化)

其他常用的后缀组成的词如:

programmable	可编程的	portable	便携的
avoidable	可以避免的	audible	听得见的
fundamental	基本的	abundant	富饶的
apparent	显然的	cultured	有文化的
useful	有用的	economical	经济的
useless	无用的	numerous	众多的
hardware	硬件	software	软件
reliability	可靠性	confidentiality	保密性

(5)借用词

借用词是指借用公共英语及日常生活用语中的词汇来表达专业含义。借用词一般来自厂商名、商标名、产品代号名、发明者名、地名等,也可将公共英语词汇演变成专业词义而实现。也有对原来词汇赋予新的意义的。如:

cache	高速缓存	semaphore	信号量
firewall	防火墙	mail bomb	邮件炸弹
fitfall	子程序入口	flag	标志,状态

英语科技文章中有很多词汇并不是专业术语,但在日常口语中用得也不是很多,它们多见于书面语中。掌握这类词对阅读科技文献或写科技论文十分重要。如:

accordance	按照	acknowledge	承认
alternative	交替的	application	应用
appropriate	恰当的	circumstance	情况
compensation	补偿	confirm	证实
modification	修改	inclusion	包括
indicate	指示	induce	导致
nonetheless	然而	nevertheless	然而

（6）通过词类转化构成新词

指一个词不变化词形，而由一种词类转化为另一种或几种词类，有时发生重音或尾音的变化。英语中名词、形容词、副词、介词可以转化成动词，动词、形容词、副词、介词可以转化成名词。但最活跃的是名词转化成动词和动词转化成名词。

如：island（名词）小岛→ island（动词）隔离
coordinate（动词）协调→ coordinate（名词）坐标
center（名词）中心 → center（动词）集中
break（动词）打破 → break（名词）间歇
close（关上） → close（副词）靠近
clear（形容词）明确的 → clear（动词）清除
all（形容词）全部的 → all（代词）全体
hard（形容词）坚硬的 → hard（副词）努力地
but（连词）但是 → but（介词）除了

1.2.2 词汇缩略

词汇缩略是指将较长的单词取其首部或主干构成与原词同义的短单词，或者将组成词汇短语的各个单词的首字母拼接为一个大写字母的字符串。通常词汇缩略在文章索引、前序、摘要、文摘、电报、说明书、商标等科技文章中频繁采用。对计算机专业来说，在程序语句、程序注释、软件文档、互联网信息、文件描述中也采用了大量的缩略词汇作为标识符、名称等。缩略词汇的出现方便了印刷、书写、速记，以及口语交流等，但也同时增加了阅读和理解的困难。

词汇缩略有以下 4 种形式。

1．节略词（Clipped words）

某些词汇在发展过程中为方便起见逐渐用它们的前几个字母来表示，这就是节略词。也有的节略词是在一个词组中取各词的一部分，重新组合成一个新词，表达的意思与原词组相同。

如：maths —— mathematics　　　　　数学
ad —— advertisement　　　　　广告
kilo —— kilogram　　　　　公斤
lab —— laboratory　　　　　实验室
radar —— radio detection and ranging　　　　　雷达
transceiver —— transistor receiver　　　　　收发机
TELESAT —— Telecommunications satellite　　　　　通信卫星

2．首字词（Initials）

首字词与缩略词基本相同，区别在于首字词必须逐字母念出。

如：CAD——Computer Aided Design（计算机辅助设计）
CPU——Central Processing Unit（中央处理器）
DBMS——Data Base Management System（数据库管理系统）
UFO——Unidentified Flying Object（不明飞行物）
CGA——Color Graphics Adapter（彩色图形适配器）

3．缩写词（Abbreviation）

缩写词并不一定由某个词组的首字母组成，有些缩写词仅由一个单词变化而来，而且大多数缩写词每个字母后都附有一个句点。

如：e.g. —— for example
Ltd. —— limited
sq. —— square

4．缩略词（Acronyms）

缩略词是指由某些词组中各个词的首字母所组成的新词。

如：ROM——Read Only Memory（只读存储器）
RAM——Random Access Memory（随机访问存储器）
RISC——Reduced Instruction Set Computer（精简指令集计算机）
CISC——Complex Instruction Set Computer（复杂指令集计算机）
COBOL——Common Business Oriented Language（面向商务的通用语言）

1.2.3　计算机专用术语与命令

在计算机语言、程序语句、程序文本注释、系统调用、命令字、保留字、指令字以及网络操作中广泛使用专业术语进行信息描述。随着计算机技术的发展，这样的专业术语还会进一步增加。

1．专用的软件名称及计算机厂商名

人类相互交流信息所用的语言称为自然语言，但是当前的计算机还不能理解自然语言，它能理解的是计算机语言，即软件。软件分成系统软件和用户软件。近几年来，随着计算机技术的发展，新的软件不断推出。下面是一些常用软件的名称。

Authorware	专业多媒体软件（属于 Aaobc 公司）
Dreamweaver	网页设计软件（属于 Aaobc 公司）
MATLAB	科学计算软件（属于 MATH WORKS 公司）
Photoshop	图像处理软件（属于 Adobe 公司）
Internet Explorer	互联网浏览器（属于 Microsoft 公司）
Java	网络编程语言（属于 Oracle 公司）
Excel	电子表格软件（属于 Microsoft 公司）

下面给出的是一些著名计算机公司的译名。

Microsoft	微软	Philip	飞利浦
Apple	苹果	DELL	戴尔
Panasonic	松下	Acer	宏碁

Intel　英特尔　　　　　　Hewlett-Packard (HP)　惠普
Samsung　三星　　　　　Epson　爱普生

2．DOS 系统

DOS（Disk Operating System）是个人计算机磁盘操作系统，DOS 是一组非常重要的程序，它帮助用户建立、管理程序和数据，也管理计算机系统的设备。DOS 是一种层次结构，包括 DOS BIOS（基本输入输出系统）、DOS 核心部分和 DOS COMMAND（命令处理程序）。

一般情况下，在 DOS 启动盘上有配置系统文件 CONFIG.SYS，在该文件内给出有关系统配置命令，能确定系统的环境。配置系统包括以下 9 个方面的内容。

（1）设置 Ctrl-Break（BREAK）检查

格式：BREAK = [ON] | [OFF]

隐含是 BREAK = OFF，这时 DOS 只对标准输出操作、标准输入操作、标准打印操作和标准辅助操作检查 Ctrl-Break.

如果设置 BREAK = ON，则 DOS 在它被调用的任何时候都检查 Ctrl-Break，比如编译程序，即使没有标准设备操作，在编译过程中遇到错误，也能使编译停止。

（2）指定磁盘缓冲区的数目（BUFFERS）

格式：BUFFERS = X，X 是 1～99 的数。

（3）指定国家码及日期时间格式（COUNTRY）

格式：COUNTRY = XXX，XXX 是电话系统使用的三数字的国际通用国家码。

（4）建立能由文件控制块打开的文件数（FCBS）

格式：FCBS = m，n

m 取值 1～255。

n 指定由 FCBS 打开但不能由 DOS 自动关闭的文件数，n 取值 0～255，约定值是 0。

（5）指定能一次打开的最大文件数（FILES）

格式：FILES = X

X 取值 8～255，约定值是 8。

（6）指定能访问的最大驱动器字母（LASTDRIVE）

格式：LASTDRIVE = X

X 可以是 A 到 Z 之间的字母，它表示 DOS 能接受的最后一个有效驱动器字母，约定值为 E。

（7）指定高层命令处理程序的文件名（SHELL）

格式：SHELL = [d：] [path] filename [.ext] [parm1] [parm2]

（8）安装驱动程序（DEVICE）

格式：DEVICE = [d：] [path] filename [.ext]

除标准设备外，如果用户增加了别的设备，就要由用户自己提供相应的驱动程序。在 CONFIG.SYS 中加上命令 DEVICE = 驱动程序名称。

（9）指定堆栈空间（STACKS）

格式：STACKS = n，s

n 是堆栈的框架个数，8～64

s 是每层堆栈框架的字节数，32～512

3．计算机专用命令和指令

程序设计语言同任何一门自然语言一样，有它自己的一套词法和语法规则，只是它的规则很少，每一条语句的规定都很严格。到目前为止，大部分的计算机语言的词汇都是取自英语词汇中一个很小的子集和最常用的数学符号。由于各个计算机指令系统所具有的功能大致相同，各个程序设计语言也大体包含了函数、过程、子程序、条件、循环以及输入和输出等部分，所以它们必然存在一些共同的词汇特点和语法特点。

系统命令与程序无关，而且语法结构简单。主要的系统命令有系统连接命令、初始化命令、程序调试命令和文件操作命令。格式为：

系统命令<CR>

其中<CR>为回车换行符，一般而言，系统命令总是立即被执行，但某些系统命令也可以用于程序执行。至于词汇特点，同一功能的命令在不同机器中一般以相同的单词表示，如删除文件命令 DELETE，列磁盘文件命令 DIR，拷贝文件命令 COPY。但有相当一部分系统命令名是由各厂家或公司自己定义的，如清屏命令就有 CLEAR、CLS、CLR、HOME 等几种。

每一个处理器都具有很多指令，每一台机器也具有很多系统命令，不同的操作系统也定义了不同的操作命令，它们通常是缩写的，牢记这些指令，就熟悉了计算机的操作；了解缩写的含义，也就了解了所用的操作的含义。例如：

创建目录　　MD（make directory）
改变目录　　CD（change directory）
删除目录　　RD（remove directory）
列表目录　　DIR（directory）
重命名　　　REN（rename）
中断请求　　INT（call to interrupt procedure）
中断返回　　IRET（interrupt return）
总线封锁命令　LOCK（assert bus lock signal）

4．网络专用术语

（1）Internet 专用缩写术语

- TCP/IP：Internet 使用的一组网络协议，其中 IP 是 Internet Protocol，即网际协议；TCP 是 Transmission Control Protocol，即传输控制协议。IP 提供基本的通信，TCP 提供应用程序所需要的其他功能。
- SMTP：Simple Mail Transfer Protocol，简单邮件传送协议。用于电子邮件传送。
- HTTP：Hypertext Transfer Protocol，超文本传输协议。用于 World Wide Web 服务。
- SNMP：Simple Network Management Protocol，简单网络管理协议。用于网络管理。
- NFS：Network File System，网络文件系统。用于实现计算机间共享文件系统。
- UDP：User Datagram Protocol，用户数据报协议。用于可靠性要求不高的场合。
- ARP：Address Resolution Protocol，地址解析协议。用于从 IP 地址找出对应的以太网地址。
- RARP：Reverse Address Resolution Protocol，逆向地址解析协议。用于从以太网地址找出对应的 IP 地址。
- ICMP：Internet Control Message Protocol，Internet 控制信息协议。
- IGMP：Internet Group Multicast Protocol，Internet 成组广播协议。

（2）Internet 服务
- E-mail：电子邮件，指通过计算机网络收发信息的服务。电子邮件是 Internet 上最普遍的应用，它加强了人与人之间沟通的渠道。
- Telnet：远程登录。用户可以通过专门的 Telnet 命令登录到一个远程计算机系统，该系统根据用户账号判断用户对本系统的使用权限。
- FTP：File Transfer Protocol，文件传输协议。利用 FTP 服务可以直接将远程系统上任何类型的文件下载到本地计算机上，或将本地文件上载到远程系统。它是实现 Internet 上软件共享的基本方式。
- Usenet：新闻组，又称网上论坛或电子公告板系统（Bulletin Board System, BBS），是人们在一起交流思想观点、公布公共注意事项和寻求帮助的地方。
- WWW：World Wide Web，万维网。当前 Internet 上最重要的服务方式。WWW 是由欧洲核子研究中心（CERN）研制的，它将位于全球 Internet 上不同地点的相关多媒体信息有机地编织在一起，称为 Web 的集合。

（3）Internet 地址
- Domain Name

域名，它是 Internet 中主机地址的一种表示方式。域名采用层次结构，每一层构成一个子域名，子域名之间用点号隔开并且从右到左逐渐具体化。域名的一般表示形式为：计算机名、网络名、机构名、一级域名。一级域名有一些规定，用于区分机构和组织的性质，如 edu 为教育机构，com 为商业单位，mil 为军事部门，gov 为政府机关，org 为其他组织。

用于区分地域的一级域名采用标准化的两个字母的代码。如：

cn	中国	ca	加拿大
us	美国	au	澳大利亚
gb	英国（官方）	uk	英国（通用）
tw	中国台湾	hk	中国香港
fr	法国	un	联合国
nz	新西兰	dk	丹麦
ch	瑞士	de	德国
jp	日本	sg	新加坡
aq	南极洲	it	意大利

- 电子邮件地址

在 Internet 上，电子邮件（E-mail）地址具有如下统一的标准格式：用户名@主机域名。例如，wang@online.sh.cn 是一个电子邮件的地址，其中 wang 是用户名，@是连接符，online.sh.cn 是"上海热线"的主机域名，这是注册"上海热线"后得到的一个 E-mail 地址。

1.3 专业英语中的常用语法知识

1.3.1 动词不定式

动词不定式是非谓语动词的一种，由不定式符号 to 加动词原形构成。之所以叫作"不定式"，是因为它的形式不像谓语动词那样受到主语人称和数的限制。但是，动词不定式又具有动词的许多特点，它可以有自己的宾语、状语及宾语补足语。动词不定式和它的宾语、状语及宾语补足语构成不定式短语。动词不定式还有时态和语态的变化,参见表1.1。

表 1.1　动词不定式用法表

时　态	语态（主动）	语态（被动）	用　　法
一般式	to do	to be done	表示动作有时与谓语动词表示的动作同时发生,有时发生在谓语动词的动作之后
进行式	to be doing	—	表示动作正在进行,或与谓语动词表示的动作同时发生
完成式	to have done	to have been done	表示动作发生在谓语动词表示的动作之前

例：Today we use computers to help us to do most of our work.

译文：如今我们用计算机帮助做大部分工作。

句中,动词不定式 to help 带有宾语 us 和宾语补足语 do most of our work。

动词不定式通常具有名词性、形容词性和副词性,因此可以充当句子的主语、表语、宾语、定语、状语和补足语。下面分别叙述动词不定式在句中的作用。

1. 作主语

动词不定式（短语）作主语,较多地用来表示一个特定的行为或事情,谓语动词需要用第三人称单数,且常用 it 作形式主语。

例：To know something about computer is important.

译文：懂得一些计算机的知识很重要。

句中,To know something about computer 是动词不定式短语,在句子中作主语。

不定式短语作主语时,为了句子的平衡,常常把它放在句尾,而用 it 作形式主语代替不定式放在句首。

例：It would perhaps be unwise to forecast undue restrictions on the nature of the ultimate achievement.

译文：在对大自然的终极成就的预测中,如加上过分的限制,可能是不太明智的。

2. 作表语

不定式可放在系动词后面作表语。如：

例：To see is to believe.

译文：眼见为实。

句中,to believe 放在 is 后面作表语。

例：Our task today is to work out the design.

译文：我们今天的工作是把设计做出来。

3. 作宾语

不定式（短语）在某些及物动词后可作宾语。这类及物动词通常有 want, like, wish, hope, begin, decide, forget, ask, learn, help, expect, intend, promise, pledge 等。

例：This helps to save coal and reduce the cost of electricity.

译文：这有助于节约用煤以及降低发电成本。

例：They decided to do the experiment again.

译文：他们决定再次做这个实验。

当某些动词后面作宾语的不定式必须有自己的补语才能使意思完整时,要用 it 作形式宾语，而将真正的宾语（即不定式）后置。常用这种结构的动词有 think、find、make、consider、feel 等。

例：The use of semiconductor devices together with integrated circuits make it possible to develop miniaturized equipment.

译文：半导体装置和集成电路一起使用使得发展微型设备成为可能。

句中，不定式短语 to develop miniaturized equipment 作宾语，it 是形式上的宾语。

4．作定语

动词不定式（短语）作定语时，通常放在它所修饰的名词（或代词）之后。

例：He never had the chance to learn computer.

译文：他从来没有学习计算机的机会。

句中，to learn computer 是动词不定式，在句中作定语，修饰和限定 the chance。

有时，动词不定式与它所修饰的名词是逻辑上的动宾关系。

例：We usually define energy as the ability to do work.

译文：我们通常将能量定义为做功的能力。

5．作状语

不定式作状语可以修饰句中的动词、形容词、副词或全句，主要表示目的、程度、结果、范围、原因等。

例：We are glad to hear that you have bought a computer.

译文：听说你买了一台计算机，我们十分高兴。（表示原因）

例：To meet our production needs, more and more electric power will be generated.

译文：为了满足生产的需要，将生产越来越多的电力。（表示目的）

例：Solar batteries have been used in satellites to produce electricity.

译文：人造卫星上已经用太阳能电池发电。（表示结果）

6．作宾语补足语

某些及物动词要求不定式作宾语补足语。宾语补足语是对宾语的补充说明。

例：A force may cause a body to move.

译文：力可以使物体移动。

句中，a body 是宾语，不定式 to move 是宾语补足语。

例：Conductors allow a large number of electrons to move freely.

译文：导体允许大量的电子自由运动。

当 make、let、have、see、hear、watch、notice、feel 等动词后面用不定式作宾语补足语时，不定式都不带 to。这一点特别重要。

例：I often hear people talk about this kind of printer.

译文：我经常听人们谈论这种打印机。

句中，talk about this kind of printer 是个不带 to 的动词不定式短语，在句中作宾语 people 的补足语。

7．作主语补足语

当主动语态的句子变成被动语态时，主动语态句子中的宾语补足语就在被动语态中变成主语补足语。若主动语态中的宾语补足语由动词不定式构成，则该句变为被动语态后它也相应地变为主语补足语。

例：He was asked to do the experiment at once.
译文：有人请他马上做实验。

但是，当 make、let、have、see、hear、watch、notice、feel 等动词的句子变为被动语态时，原来在主动语态时作宾语补足语的动词不定式这时也变为主语补足语，此时，动词不定式中的 to 不能省略。

例：He was made to finish repairing the printer.
译文：他被迫马上修好打印机。

8．不定式的特殊句型
例：I am afraid the box is too heavy for you to carry it.（too...to 太……以至于……）
译文：这箱子太重，恐怕你搬不动。

1.3.2 分词

分词是非谓语动词的一种。分词有现在分词和过去分词两种。规则动词的现在分词由动词原形加-ing 构成，过去分词由动词原形加-ed 构成；不规则动词的分词形式，其构成是不规则的。分词没有人称和数的变化，具有形容词和副词的作用，同时还保留着动词的特征，只是在句中不能独立作谓语。

现在分词所表示的动作具有主动的意义，而及物动词的过去分词表示的动作具有被动的意义。现在分词与过去分词在时间关系上，前者表示动作正在进行，后者表示的动作往往已经完成。现在分词表示的动作与谓语动词表示的动作相比，具有同时性，而过去分词则具有先时性。分词在各种时态、语态下的表现形式如表 1.2 所示。

表 1.2　分词用法表

时态	与主语动词同时	doing
	先于主语动词	having done
语态	表示主动	现在分词 doing
	表示被动	过去分词 done
	表示动作已经发生	不及物动词的过去分词

分词在句子中具有形容词词性和副词词性，可以充当句子的定语、表语、状语和补足语。下面分别举例说明现在分词和过去分词在句子中的作用。

1．现在分词
（1）作定语
例：They insisted upon their device being tested under operating conditions.
译文：他们坚持他们的装置要在运转条件下检测。
（2）作表语
例：The result of the experiment was encouraging.
译文：实验结果令人鼓舞。
（3）作补足语
例：You'd better start the computer running.
译文：你还是把计算机启动起来好。

（4）作状语

例：While making an experiment on an electric circuit, they learned of an important electricity law.
译文：他们在做电路实验时，学到了一条重要的电学定律。

2．过去分词

（1）作定语

例：The charged capacitor behaves as a secondary battery.
译文：充了电的电容就像一个蓄电池一样。

（2）作表语

例：Some substances remain practically unchanged when heated.
译文：有几种物质受热时几乎没有变化。

（3）作补足语

例：I don't know if we can get the computer repaired in time.
译文：我不知道我们能否按时修好计算机。

（4）作状语

例：Given the voltage and current, we can determine the resistance.
译文：已知电压和电流，我们就可以求出电阻。

1.3.3 动名词

动名词是一种非谓语动词，由动词原形加词尾-ing 构成，形式上和现在分词相同。由于动名词和现在分词的形成历史、意义和作用都不一样，通常把它们看作是两种不同的非谓语动词。它没有人称和数的变化。动名词具有动词词性和名词词性，因而又可以把它称为"动词化的名词"和"名词化的动词"，在句中充当主语、表语、定语和宾语等。动名词也可以有自己的宾语和状语，构成动名词短语。动名词在各种时态、语态下的形式如表 1.3 所示。

表 1.3 动名词用法表

	时态/语态	主动	被动
动名词	一般式	doing	being done
	完成式	having done	having been done

下面分别举例说明动名词在句子中的作用。

1．作主语

动名词作主语表示一件事或一个行为，其谓语动词用第三人称单数。

例：Changing resistance is a method for controlling the flow of the current.
译文：改变电阻是控制电流流动的一种方法。

动名词作主语时，也可用 it 作形式主语，放在句首，而将真正的主语——动名词短语放在谓语之后。

例：It's no good using this kind of material.
译文：采用这类材料是毫无用处的。

2．作宾语

动名词可以在一些及物动词和介词后作介词宾语。要求动名词作宾语的常用的及物动词有 finish、enjoy、avoid、stop、need、start、mean 等。

例：This printer needs repairing.

译文：这台打印机需要修理一下。

例：I remember having repaired this machine.

译文：我记得曾经修过这部机器。

英语中，suggest、finish、avoid、stop、admit、keep、require、postpone、practice、fancy、deny 等动词都用动名词作宾语，不能用不定式作宾语。但是在 love、like、hate、begin、start、continue、remember、forget、regret 等词后面既可以用动名词作宾语，也可以用动词不定式作宾语。

例：Do you like watching/to watch TV？

译文：你喜欢看电视吗？

动名词作宾语时，如本身带有补足语，则常用 it 作形式宾语，而将真正的宾语——动名词放在补足语的后面。

例：I found it useless arguing with her.

译文：我发现与她辩论没有用。

3．作表语

动名词作表语为名词性表语。表示主语的内容，而不说明主语的性质。主语常为具有一定内涵的名词，这点与不定式作表语相似。动名词作表语与动词的进行时的区别在于主语能否执行该词的行为，能执行，即为进行时；否则，即为动名词作表语（系表结构）。

例：The function of a capacitor is storing electricity.

译文：电容器的功能是存储电能。

句中的"storing"是动名词，"storing electricity"作表语。

例：Seeing is believing.

译文：眼见为实。

句中，动名词 Seeing 作主语，believing 作表语。

4．作定语

动名词作定语为名词性定语，说明名词的用途，与所修饰名词之间没有逻辑主谓关系，这点是与现在分词作定语相区别的关键。动名词作定语只能使用单词，不可用动名词短语；只能放在所修饰名词前面，不可后置。

例：English is one of the working languages at international meeting.

译文：英语是国际会议上使用的工作语言之一。

5．作宾语补足语

动名词在句中的作用相当于名词，故可作宾语补足语。动名词只能在少数动词后作宾语补足语，补充说明宾语的性质、行为或状态，与宾语具有逻辑主谓关系。

例：We call this process testing.

译文：我们称这个过程为检测。

句中，动名词 testing 作宾语 this process 的补足语。

1.3.4　被动语态

语态是动词的一种形式，它表示主语和谓语的不同关系。语态有两种：主动语态和被动语态。主动语态表示句子的主语是谓语动作的发出者；被动语态表示主语是谓语动作的承受者。也就是说，主动语态句子中的宾语，在被动语态中作句子的主语。由于被动语态句子的主语是谓语动作的承受者，故只有及物动词才会有被动语态。

主动语态：He designed this building.

他设计了这座大楼。

被动语态：This building was designed by him.

这座大楼是他设计的。

在科技英语中，为了着重说明客观事物和过程，就会更多的用到被动语态。被动语态构成如下：

主语＋be＋（及物动词）过去分词

1．科技英语中主要时态的被动语态形式

（1）一般现在时

一般现在时的被动语态构成如下：

主语＋am（is，are）+及物动词的过去分词

例：I am asked to solve this problem by him.

译文：他请我解决这个问题。

例：The switches are used for the opening and closing of electrical circuits.

译文：开关是用来开启和关闭电路的。

（2）一般过去时

一般过去时的被动语态构成如下：

主语＋was（were）+及物动词的过去分词

例：That plotter was not bought in Beijing.

译文：那台绘图仪不是在北京买的。

例：The insulator was burned out by overheating.

译文：绝缘体因过热而被烧毁。

（3）一般将来时

一般将来时的被动语态构成如下：

主语＋will be＋及物动词的过去分词

当主语是第一人称时，可用：

主语＋shall be＋及物动词的过去分词

例：I shall not be allowed to do it.

译文：不会让我做这件事的。

例：What tools will be needed for the job?

译文：工作中需要什么工具？

（4）现在进行时

现在进行时的被动语态构成如下：

主语＋is（are）being＋及物动词的过去分词

例：Our printer is being repaired by John.

译文：约翰正在修理我们的打印机。

例：Electron tubes are found in various old products and are still being used in the circuit of some new products.

译文：在各种老产品里看到的电子管，在一些新产品的电路中也还在使用。

（5）过去进行时

过去进行时的被动语态构成如下：

主语 + was（were）being + 及物动词的过去分词

例：The laboratory building was being built then.

译文：实验大楼当时正在建造。

（6）现在完成时

现在完成时的被动语态构成如下：

主语 + have（has）been + 及物动词的过去分词

例：New techniques have been developed by the research department.

译文：研究部门研发了新技术。

（7）过去完成时

过去完成时的被动语态构成如下：

主语 + had been + 及物动词的过去分词

例：Electricity had been discovered for more than one thousand years by the time it came into practical use.

译文：电在发现一千多年之后，才得到实际应用。

2．常用被动语态的几种情况

（1）当我们强调的是动作的承受者或给动作的承受者较大关注时，多用被动语态。这时，由于动作的执行者处于次要地位，句子中 by 引导的短语可以省略。

例：The virus in the computer has been found out.

译文：计算机中的病毒已经找出来了。

（2）当我们不知道或不想说出动作的执行者时，可使用被动语态。这时句子中不带由 by 引导的短语。

例：Electricity was discovered long ago.

译文：电是很久以前发现的。

（3）当动作的执行者是"物"而不是"人"时，常用被动语态。

例：This machine is controlled by a computer.

译文：这台机器由计算机控制。

（4）当动作的执行者已为大家所熟知，而没有必要说出来时，也常常使用被动语态。

例：This factory was built twenty years ago.

译文：这座工厂是20年前兴建的。

（5）使用被动语态能更好地安排句子。

例：The professor came into the hall and was warmly applauded by the audience.

译文：教授走进大厅，大家热烈鼓掌。

1.3.5 定语从句

定语从句又称关系从句，在句子中起定语作用，修饰一个名词或代词，有时也可修饰一个句子。被定语从句修饰的名词、词组或代词叫先行词，定语从句通常跟在先行词的后面。

例：This is the software that I would like to buy.

译文：这就是我想买的那个软件。

that I would like to buy 是定语从句，the software 是先行词。

通常，定语从句都由关系代词 that、which、who、whom、whose 和关系副词 when、where、why、how 引导。关系代词和关系副词往往放在先行词和定语从句之间，起联系作用，同时还代替先行词在句中担任一定的语法成分，如主语、宾语、定语和状语等。

例：The man who will give us a lecture is a famous professor.

译文：将要给我们做讲演的人是位著名的教授。

该句中，who will give us a lecture 是由关系代词 who 引导的定语从句，修饰先行词 the man，who 在从句中作主语。

定语从句根据其与先行词的密切程度可分为限定性定语从句和非限定性定语从句。

1．限定性定语从句

限定性定语从句与先行词关系密切，是整个句子不可缺少的部分，没有它，句子的意思就不完整或不明确。这种定语从句与主句之间不用逗号隔开，译成汉语时，一般先译定语从句，再译先行词。

限定性定语从句如果修饰人，一般用关系代词 who，有时也用 that。若关系代词在句子中作主语，则 who 用得较多，且不可省略；若关系代词在句子中作宾语，就应当使用宾格 whom 或 that，但在大多数情况下都可省略。若表示所属，就应用 whose。

限定性定语从句如果修饰物，用 that 较多，也可用 which。他们可在句中作主语，也可作宾语。若作宾语，则大多可省略。

例：Those who agree with me please put up your hands.

译文：同意我的观点的人请举手。

who agree with me 是定语从句，修饰 Those。who 既是引导词，又在句中作主语，who 不能省略。

例：PCTOOLS are tools whose functions are very advanced.

译文：PCTOOLS 是功能很先进的工具。

因为 functions 和 tools 之间是所属关系，故用所有格 whose。

例：Mouse is an instrument which operators often use.

译文：鼠标是操作员经常使用的一种工具。

which 引导的定语从句修饰 an instrument。因为 which 在从句中作 use 的宾语，故可省略。

2．非限定性定语从句

非限定性定语从句与先行词的关系比较松散，从句只对先行词附加说明，如果缺少，不会影响句子的主要意思。从句与主句之间常用逗号隔开，译成汉语时，从句常单独译成一句。

非限定性定语从句在修饰人时用 who、whom 或 whose，修饰物时用 which，修饰地点和时间时用 where 和 when 引导。关系代词 that 和关系副词 why 不能引导非限定性定语从句。

例：We do experiments with a computer, which helps to do many things.

译文：我们利用计算机做实验，计算机可帮助做许多工作。

Which 引导的非限定性定语从句是对先行词 a computer 的说明。

例：The meeting will be put off till next week, when we shall have made all the preparations.

译文：会议将推迟到下周，那时我们将做好一切准备。

例：Mechanical energy is changed into electric energy, which in turn is changed into mechanical energy.

译文：机械能转变为电能，而电能又转变为机械能。

1.3.6 状语从句

英语中的状语从句通常由从属连词和起连词作用的词组来引导，用来修饰主句中的动词、形容词、副词等。

状语从句可位于主句前，也可位于主句后；前置时，从句后常用逗号与主句隔开；后置时，从句前通常不使用逗号。状语从句在句子中作状语，可表示时间、原因、目的、结果、条件、比较、方式、让步和地点等不同含义。

1．状语从句的分类

（1）时间状语从句

引导时间状语从句的连词或词组很多，但可根据所表示时间的长短以及与主句谓语动词行为发生的先后这两点去理解和区别。

这些连词或词组有：when（当……时候），as（当……时候，随着，一边……一边），while（在……期间），before（在……之前），after（在……之后），since（自从……以来），until（till）（直到……才），as soon as（一……就），no sooner…than（刚一……就……），once（一旦），every time（每次）等。

例：It changes speed and direction when it moves.

译文：在运动时它改变速度和方向。

例：Check the circuit before you begin the experiment.

译文：检查好线路再开始做实验。

（2）原因状语从句

引导原因状语从句的连词和词组有：because（因为），as（由于），since（既然，由于），now that（既然），in that（因为）等。其中前3个较常用，它们表示原因的正式程度依次为 because > since ≥ as。当原因是显而易见的或已为人们所知时，就用 as 或 since。由 because 引导的从句如果放在句首，且前面有逗号，则可以用 for 来代替。但如果不是说明直接原因，而是多种情况加以推断，就只能用 for。

例：Electric energy is used most widely mainly because it can be easily produced, controlled, and transmitted.

译文：电能用得最广，主要是因为发电容易，而且控制和输送也方便。

（3）目的状语从句

目的状语从句由 in order that（为了，以便），so that（为了，以便），that（为了），lest（以免，以防），for fear that（以免，以防）等引导。

例：He handled the instrument with care for fear that it should be damaged.

译文：他小心地弄那仪器，生怕把它弄坏。

例：You must speak louder so that you can be heard by all.

译文：你必须说大声点以便大家都能听到。

（4）结果状语从句

引导结果状语从句的连词有：so that（结果，以致），so … that（如此……以致），such …

that（这样的……以致）等。注意 so 后接形容词或副词，而 such 后跟名词。so 还可以与表示数量的形容词 many、few、much 及 little 连用，形成固定搭配。

例：This problem is so difficult that it will take us a lot of time to work it out.

译文：这道题很难，我们要用很长时间才能解出。

（5）条件状语从句

条件状语从句用来表示前提和条件。通常由以下连词引导：

if（如果），unless（除非），provided / providing that（假如），as long as（只要），in case（如果），on condition that（条件是……），suppose / supposing（假如）等。

例：A physical body will not tend to expand unless it is heated.

译文：除非受热，否则物体不会有膨胀的倾向。

（6）比较状语从句

比较状语从句经常是省略句，一般都是省略了重复部分；省略之后不影响句意，反而结构简练。部分比较状语从句还有倒装现象。

比较状语从句由下列连词引导：as … as（像……一样），than（比），not so (as) … as（不像……一样），the more … the more（愈……愈），as … so（正如……那样）等。

例：Electron tubes are not so light in weight as semiconductor devices.

译文：电子管的重量不如半导体器件那么轻。

例：He finished the work earlier than we had expected.

译文：他完成这项工作比我们预计的要早。

（7）方式状语从句

方式状语从句通常由 as（如同，就像），as if（as though）（好像，仿佛）等连词引导。

as 引导的方式状语从句常常是一个省略句。as if 和 as though 两者的意义和用法相同，引出的状语从句常是一个虚拟语气的句子，表示没有把握的推测，或是一种夸张的比喻。(just) as …so…引导的方式状语从句通常位于主句后，意为"正如……，就像"。

例：The earth itself behaves as though it were an enormous magnet.

译文：地球本身的作用就像一个大磁铁一样。

例：They completely ignore these facts as if they never existed.

译文：他们完全不理会这些事实就好像它们不存在一样。

（8）让步状语从句

让步状语从句表示在相反的（不利的）条件下，主句行为依然发生了。

引导让步状语从句的有：(al)though（虽然），even if(though)（即使），as（尽管），whatever（不管），however（无论怎样），no matter（how, what, where, when）（不管怎样，什么，哪里，何时），whether…or（不论……还是）等。

例：It is important to detect such flows, even if they are very slight, before the part is installed.

译文：在安装部件之前，即使变形很轻微，也必须探测出来。

例：Much as computer languages differ, they have something in common.

译文：尽管计算机语言之间各不相同，但它们仍有某些共同点。

（9）地点状语从句

引导地点状语从句的词有：where(在……地方, 哪里), wherever(在任何地方), everywhere（每一……地方）等。

例：She found her pen where she had left it.

译文：她的笔是她在原来放笔的地方找到的。

2．状语从句的翻译方法

状语从句常用的翻译方法有顺译法、倒译法、转译法和缩译法。

（1）顺译法

一般的句子可以按照原文提供的顺序直接翻译。当表示目的、原因等的状语从句在主句之前出现时，直接按照原句语序翻译。如果这些状语出现在主句之后，可以将它们提前或者保持原句顺序，翻译在主句之后，对主句意思起到补充说明的作用。

例：Whenever you need any specific information, you can search it by Internet.

译文：每当你需要任何专业信息时，你都可以通过互联网搜索得到。

例：The Internet is so powerful that you can get various information through it.

译文：互联网是如此强大，以至于你可以通过它获取各种各样的信息。

（2）倒译法

当原文中的时间状语和地点状语在主句后面时，必须倒译；当原文中的原因状语从句、条件状语从句和让步状语从句在主句后面时，一般也可以倒译；另外当特殊比较从句在主句后面时，必须倒译。

例：Many businesses became aware of network when they bought an expensive laser printer and wanted all the PCs to print to it.

译文：当企业购买了一台昂贵的激光打印机，并希望其所有的计算机都能使用该打印机时，他们就想到了网络。

（3）转译法

当通过对原文的逻辑含义进行分析后，会发现 when、where 不再单纯地表示时间、地点，或者翻译成"当……"或"在……"不合适时，可以考虑这些词也可以表示"如果"的意思。另外，当状语从句比较短，而关联词可以省略时，可以把状语从句翻译成并列成分，这样也使得句子比较紧凑。

例：Where the Hz is too small a unit, we may use the MHz.

译文：当用赫兹作单位太小时，我们可以使用兆赫兹。

例：Our whole physics universe, when reduced to the simplest terms, is made up of two things: energy and matter.

译文：我们的整个物理世界，如果用简单的话来说，是由能量和物质这两样东西组成的。

（4）缩译法

有些关联词，比如"so...that..."在很多情况下，可以省略翻译，这样使得汉语的译文就很简练，对于这样的句子就可以采用"缩译法"进行翻译。

例：Computers work so fast that they can solve a very difficult problem in a few seconds.

译文：计算机工作如此迅速，一个很难的题目几秒钟内就可以解决。

1.3.7 时态简介

1．一般现在时

一般现在时用于如下情况：

- 经常性或习惯性的动作，常与表示频度的时间状语连用。时间状语有：every …，sometimes，at …，on Sunday 等。
- 表示客观真理、客观存在或科学事实。
- 表示格言或警句，此用法如果出现在宾语从句中，即使主句是过去时，从句也要用一般现在时。
- 现在时刻的状态、能力、性格和个性等。

例：I go to school every day.

译文：我每天去上学。

例：The adult mosquito usually lives for about thirty days, although the life span varies widely with temperature, humidity, and other factors of the environment.

译文：成年蚊子通常可活约 30 天，尽管其寿命的长短还随温度、湿度和其他环境因素而变化很大。

2．一般过去时

一般过去时用于如下情况：
- 在确定的过去时间里所发生的动作或存在的状态。时间状语有：yesterday、last week、an hour ago、the other day、in 1999、before 等。
- 表示在过去一段时间里，经常性或习惯性的动作。
- wish，wonder，think，hope 等用过去时，作试探性的询问、请求、建议等。
- 用过去时表示现在，表示委婉语气。有动词 want、hope、wonder、think、intend，情态动词 could，would。

例：How many subjects did you study last term?

译文：上学期你学了多少门课程？

例：I used to enjoy gardening, but I don't like it any more.

译文：我曾经很喜欢园艺，现在一点儿也不喜欢了。

3．一般将来时

一般将来时的结构为：will（shall）+ 动词原形，可用于如下情况：
- 一般将来时的时间状语有 shall 和 will。shall 用于第一人称，常被 will 所代替，will 在陈述句中用于各人称，在征求意见时常用于第二人称。
- be going to + 不定式，表示将来。
- be + 不定式表示将来，按计划或正式安排将发生的事。
- be about to + 不定式，意为马上做某事。但要注意：be about to 不能与 tomorrow，next week 等表示明确将来时的时间状语连用。

例：I am going to travel around the world.

译文：我计划周游世界。

例：Some day people will go to the moon.

译文：总有一天，人们会到月球上去的。

4．现在进行时

现在进行时的结构为：be + 现在分词，可用于如下情况。
- 表示现在（指说话人说话时）正在发生的事情。
- 习惯进行：表示长期的或重复性的动作，说话时动作未必正在进行。

- 表示渐变的动词有：get、grow、become、turn、run、go、begin 等。
- 与 always、constantly、forever 等词连用，表示反复发生的动作或持续存在的状态，往往带有说话人的主观色彩。

例：I am speaking with him on the phone.
译文：我正在和他通电话。
例：What are you studying?
译文：你在学什么？

5．过去进行时

过去进行时的结构为：were（was）+ 现在分词，可用于如下情况。
- 过去进行时的主要用法是描述一件事发生的背景。常用的时间状语有：this morning、the whole morning、all day yesterday、from nine to ten last evening、when 及 while 等。
- 一个长动作发生的时候，另一个短动作发生。

例：What were you doing at nine last night?
译文：昨晚9点时你在做什么？
例：Some students were playing football, while others were running around the track.
译文：一些学生在踢足球，同时另一些学生正在跑道上跑步。

6．将来进行时

将来进行时的结构为：will/be +现在分词，可用于如下情况。
- 表示将来某时进行的状态或动作，常用的时间状语有：soon、tomorrow、on Sunday、by this time、this evening、in two days 及 tomorrow evening 等。
- 按预测将来会发生的事情。

例：He will be playing football soon.
译文：他一会将去踢足球。

7．现在完成时

现在完成时的构成为：have（has）+ 过去分词，可用于如下情况。
- 用来表示之前已发生或完成的动作或状态，其结果的确和现在有联系。常用的时间状语有：for、since、so far、ever、never、just、yet、till/until、up to now、in past years 及 always 等。
- 动作或状态发生在过去，但它的影响现在还存在，强调过去的事情对现在的影响。
- 也可表示持续到现在的动作或状态，动词一般是延续性的，如 live、teach、learn、work、study 及 know 等。

例：I have visited your school before.
译文：我以前曾经去过你们学校。
例：I have finished my homework now.
译文：现在我已经做完作业了。

8．过去完成时

过去完成时的构成形式是 had + 过去分词，可用于如下情况。
- 表示过去的过去，常用的时间状语有 before、by、until、when、after、once 及 as soon as 等。
- 在 told、said、knew、heard、thought 等动词后的宾语从句。
- 状语从句：在过去不同时间发生的两个动作中，发生在先，用过去完成时；发生在后，用一般过去时。
- 表示意向的动词，如 hope、wish、expect、think、intend、mean 及 suppose 等，用过去完成时表示原本……，未能……

例：He said that he had never been to Paris.
译文：他说他从未去过巴黎。
例：I did not know what he had done it for?
译文：我以前不知道他做这个究竟为了什么？

9．将来完成时

将来完成时的结构是：will have + 过去分词，可用于如下情况。
- 表示某事持续到将来某一时为止一直有的状态。
- 表示将来某一时或另一个将来的动作之前，已经完成的动作或获得的经验。

例：Mary has been to Sumatra and Iran as well as all of Europe. By the time she is twenty, she will have been to almost everywhere.
译文：玛丽到过苏门答腊、伊朗以及整个欧洲。到她 20 岁的时候她几乎将去过任何地方。

1.4 计算机专业资料的翻译与写作

1.4.1 专业英语的阅读

阅读实际上就是语言知识、语言技能和智力的综合运用。在阅读过程中，这三个方面的作用总是浑然一体，相辅相成的。词汇和语法结构是阅读所必备的语言知识，但仅此是难以进行有效阅读的，学生还需具备运用这些语言知识的能力，包括根据上下文来确定准确词义和猜测生词词义的能力，辨认主题和细节的能力，正确理解连贯的句与句之间、段与段之间的逻辑关系的能力。这里所指的智力是学生的认知能力，包括记忆、判断和推理的能力。因为在阅读专业英语文章中常常有一些要求领悟文章的言外之意和作者的态度、倾向方面的内容。

阅读能力的提高是由多方面因素决定的，学生应从以下三个方面进行训练。

（1）打好语言基本功

扎实的语言基础是提高阅读能力的先决条件。首先，词汇是语言的建筑材料。提高专业英语资料的阅读能力必须扩大词汇量，尤其是掌握一定量的计算机专业词汇。如词汇量掌握得不够，阅读时就会感到生词多，不但影响阅读的速度，而且影响理解的程度，因此不能进行有效的阅读。其次，语法是语言中的结构关系，用一定的规则把词或短语组织到句子中，以表达一定的思想。熟练掌握英语语法和惯用法也是阅读理解的基础。在阅读理解中必须运用语法知识来辨认出正确的语法关系。如果语法基础知识掌握得不牢固，在阅读中遇到结构复杂的难句、长句，就会不知所措。

（2）在阅读实践中提高阅读能力

阅读能力的提高离不开阅读实践。在打好语言基本功的基础上，还要进行大量的阅读实践。词汇量和阅读能力的提高是一种辩证关系：要想读得懂，读得快，就必须扩大词汇量；反之，要想扩大词汇量，就必须大量阅读。同样，语法和阅读之间的关系也是如此：有了牢固的语法知识能够促进阅读的顺利进行，提高阅读的速度和准确率；反之，通过大量的阅读实践又能够巩固已掌握的语法知识。只有在大量的阅读中，才能培养语感，掌握正确的阅读方法，提高阅读理解能力。同时在大量的阅读中，还能巩固计算机专业知识以及了解到计算机专业的发展趋势，对于跟踪计算机技术的发展很有好处。

（3）掌握正确的阅读方法

阅读时，注意每次视线的停顿应以一个意群为单位，而不应以一个单词为单位。若一个单词一个单词地读，当读完一个句子或一个段落时，前面读的是什么内容早就忘记了。这样读不仅速度慢，还影响理解。因此，正确的阅读方法可以提高阅读速度同时提高阅读理解能力。

常用的有效的阅读方法有三种，即略读（Skimming）、查读（Scanning）和精读（Reading for full understanding）。

略读是指以尽可能快的速度进行阅读，了解文章的主旨和大意，对文章的结构和内容获得总的概念和印象。一般地说，400字左右的短文要求在6~8分钟内完成。进行略读时精力必须特别集中，还要注意文中各细节分布的情况。略读过程中，学生不必去读细节，遇到个别生词及难懂的语法结构也应略而不读。不要逐词逐句读，力求一目数行而能知大体含义。略读时主要注意以下几点：

- 注意短文的开头句和结尾句，力求抓住文章的主旨大意。
- 注意文章的体裁和写作特点，了解文章结构。
- 注意了解文章的主题句及结论句。
- 注意支持主题句或中心思想的信息句，其他细节可以不读。

查读的目的主要是有目的地去找出文章中某些特定的信息，也就是说，在对文章有所了解的基础上，在文章中查找与某一问题、某一观点或某一单词有关的信息。查读时，要以很快的速度扫视文章，确定所查询的信息范围，注意所查信息的特点。如有关日期、专业词汇、某个事件、某个数字、某种观点等，寻找与此相关的关键词或关键段落。注意与所查信息无关的内容可以略过。

精读是指仔细地阅读，力求对文章有深层次的理解，以获得具体的信息。包括理解衬托主题句的细节，根据作者的意图和中心思想进行推论，根据上下文猜测词义等。对难句、长句，要借助语法知识进行分析，准确理解。

总之，要想提高阅读理解能力，必须掌握以下6项基本的阅读技能。

- 掌握所读材料的主旨大意。
- 了解阐述主旨的事实和细节。
- 根据上下文判断某些词汇和短语的含义。
- 既理解个别句子的意义，也理解上下文之间的逻辑关系。
- 根据所读材料进行一定的判断、推理和引申。
- 领会作者的观点、意图和态度。

没有习惯于阅读科技文章的学生常常会觉得阅读之后还总是停留在文字表面意义的理解上，而弄不清作者究竟要谈什么，形不成一个总的、清晰的概念。科技文章的内容是浓缩的，在任何一页上都会发现大量的信息。可以把一篇文章分成几个小部分、几页来读，甚至一次就读一个自然段。在阅读过程中不要忘了经常复习所读过的段落。

在阅读计算机科技文章时，还要注意在阅读中及时获取细节信息。在所有的文章中，作者都使用细节或事实来表达和支持他们的观点。阅读要想有效果，就要能够辨认并记住文章中重要的细节。一个细节就是一个段落中的一条信息或一个事实。它们或者给段落的主题提供证据，或者为其提供例子。有些细节或事实是完整的句子，而有一些只是简单的短语。在很多情况下，还必须能区分哪些是重要细节，哪些是次要细节。记住所有的细节是不可能的，但是在阅读过程中要尽量发现重要的细节并记住它们。

1.4.2 专业英语翻译

翻译是把一种语言表达的思想用另一种语言再现出来的活动。翻译要遵循两个基本原则：忠实原文和表达规范。忠实原文，就是要准确完整地表达原文的内容，使译文在表达思想、精神、风格和体裁方面起到与原文完全相同的作用。另外，译文语言必须符合规范，翻译时要使用简洁、易懂、符合汉语表达习惯的文字。要使用民族的、科学的、大众的语言，力求通顺易懂，不应有文理不通、逐字死译和生硬晦涩等现象。

翻译不是原文的翻版或者复制，从某种意义上来说是原文的再创作。其目的是使不懂原文的读者能够了解原文所表达的科技内容。科技文章并不要求像文艺作品那样形象化和具有感染性，但也必须文理通顺。这就是专业英语翻译的标准。

专业英语翻译需要熟悉和掌握如下的知识：
- 掌握一定的词汇量；
- 具备科技知识，熟悉所翻译专业；
- 了解中西方文化背景的异同，掌握科技英语中词汇的特殊含义；
- 掌握日常语言和文本的表达方式以及科技英语翻译技巧。

1. 专业术语的翻译

如果把一种语言的所有词汇作为一个词汇总集来看待，则各种词汇的分布情况和运用频率是不一样的。在词汇的总体分布中，有些词属于语言的共核部分，如功能词和日常用词。这些词构成了语言的基础词汇。此外，各个学术领域的技术术语和行业词构成了词汇总集的外缘，而处于基础词汇和外缘之间的是那些准技术词汇（sub-technical words）。各个学科领域还存在大量的行业专用表达方法和词汇，正是这些词汇在双语翻译中构成了真正的难度。在科技翻译中，准确是第一要素；如果为追求译文的流畅而牺牲准确，不但会造成科技信息的丢失，影响文献交流，还可能引起误解，造成严重后果。

由于准技术词汇的扩展意义（即技术意义）在科技文献中出现的频率要比其一般意义大得多，因此对这种词的翻译应充分考虑其技术语境，不能望词生义，翻译为一般意义。有些语境提供的暗示并不充分，这要靠译者从更广阔的篇章语境中去寻找合理的解释，消除歧义。如 table 和 form 这两个词在普通英语中都翻译为"表、表格"，但在 Access 数据库中，二者的界限是严格区分的，"table"是指数据库中的子表格，而"form"则指用于简化数据输入和查询的窗体。如果像某些译文那样把两个词都译为"表"，难免会引起译文失真，造成阅读困难。

随着社会的进步和科技的发展，新的发明创造不断涌现，随之也就出现了描述这些事物的新术语，很多新的术语是无法在现有的词典中找到准确的译文的。这就造成了在科技术语翻译上的难度，但这并不意味着我们就束手无策了，常用翻译词汇的方法有以下几种。

（1）意译

意译就是对原词所表达的具体事物和概念进行仔细推敲，以准确译出该词的科学概念。这种译法最为普遍，科技术语在可能情况下应尽量采用意译法。采用这种方法便于读者顾名思义，不加说明就能直接理解新术语的确切含义。

例：

loudspeaker	扬声器
semiconductor	半导体
videophone	可视电话
copytron = copy + electron	电子复写（技术）
E-mail = Electronic mail	电子邮件
modem = modulator + demodulator	调制解调器

（2）音译

音译就是根据英语单词的发音译成读音与原词大致相同的汉字。一般地，表示计量单位的词和一些新发明的材料或产品，它们的汉语名称在刚开始时基本就是音译的。此外，当由于某些原因不便采用意译法时，可采用音译法或部分音译法。

例：Radar 是取 radio detection and ranging 等词的部分字母拼成的，如译成"无线电探测距离设备"，会显得十分啰嗦，故采用音译法，译成"雷达"。

又如：

bit	比特（二进制信息单位）
baud	波特（发报速率单位）
quark	夸克（基本粒子，属新材料类）
nylon	尼龙（新材料类）
hertz	赫兹（频率单位）

（3）形译

用英语字母的形状形象地来表达原义，也称为"象译"。此外，科技文献常涉及到型号、牌号、商标名称及代表某种概念的字母，这些一般不必译出，直接引用即可。对于人名及公司名等名称类的词汇，翻译时可直接使用原文。

例：

I-shaped	工字形
T square	丁字尺
C network	C 形网络
X ray	X 射线
N region	N 区
Y-connection	Y 形连接
Q band	Q 波段（指 8mm 波段，频率为 36kHz～46kHz）
p-n-p junction	p-n-p 结（指空穴导电型—电子导电型—空穴导电型的结）

（4）计算机专业英语中技术词汇的翻译

科技的发展创造了大量的新词汇。这些词往往通过复合构词（compounding）或缩略表达全新的概念。我们必须跟踪计算机行业的发展，掌握新出现的词汇所表示的含义及其译名。表 1.4 列出了计算机专业的几个技术词汇。

表 1.4 专业词汇表

专业词汇	功能及含义
Internet	由多个网络互连而成的计算机网，网络互连需要遵循 TCP/IP，可译为"互联网"
.Net	是 Microsoft XML Web 服务平台。XML Web 服务允许应用程序通过 Internet 进行通信和共享数据，而不管所采用的是哪种操作系统、设备或编程语言。Microsoft .Net 平台提供创建 XML Web 服务之所需
XML	与 HTML（超文本标记语言）相比，XML 具有更强的兼容性和可扩展性，更适用于电子商务的发展，可译为"可扩展标记语言"
Tablet PC	平板电脑，专为依赖笔记本电脑、记事本、手持设备完成工作的移动电脑用户设计，是帮助用户远离工作台办公的强大工具
Macro	宏，是一系列命令和指令，这些命令和指令组合在一起，形成了一个单独的命令，以实现任务执行的自动化
Virtual reality	可译为"虚拟现实"，目前没有一个公认的虚拟现实定义，但其主要特征是以人为核心，使人身临其境，并能进行相互交流、实时操作，有如在真实世界中的感觉
MPEG	一种压缩比率较大的活动图像和声音的压缩标准
Datagram	自带寻址信息，独立从数据源传输到终点的数据包
MFC	是一个为基于 Windows 应用程序开发和创建 ActiveX 控制而建立的框架
ATL	是 ActiveX Template Library 的缩写，可译为"ActiveX 模板数据库"，它是一个专门为创建小而快的 ActiveX 组件而设计的新型数据库
GUI	译为"图形用户界面"，它来源于很有声望的施乐公司研究中心。后来，随着大众化的以 GUI 操作系统和应用为特征的 Macintosh 个人计算机的投入运行，苹果计算机把这一概念变成了商业上的成功。到 1992 年，当 Windows 3.1 成为大多数计算机上的标准发行时，GUI 才在 PC 上取代了传统的命令行界面而流行
SCSI	译为"小型计算机系统接口"，是当前使用最广泛的一种接口方式

2．翻译的过程

科技文献主要为叙事说理，其特点一般是平铺直叙，结构严密，逻辑性强，公式、数据和专业术语繁多，所以专业英语的翻译应特别强调"明确""通顺"和"简练"。所谓明确，就是要在技术内容上准确无误地表达原文的含义，做到概念清楚，逻辑正确，公式、数据准确无误，术语符合专业要求，不应有模糊不清、模棱两可之处。专业科技文献中一个概念、一个数据翻译不准将会带来严重的后果，甚至有巨大的经济损失。通顺的要求不但指选词造句应该正确，而且译文的语气表达也应正确无误，尤其是要恰当地表达出原文的语气、情态、语态、时态以及所强调的重点。简练就是要求译文尽可能简短、精练，没有冗词废字，在明确、通顺的基础上力求简洁明快、精练流畅。

翻译不能逐词逐字直译，必须在保持原意的基础上灵活引申，翻译的过程大致可分为理解、表达、校对三个阶段。

（1）理解阶段

透彻理解原著是确切表达的前提。理解原文必须从整体出发，不能孤立地看待一词一句。每种语言几乎都存在着一词多义的现象，因此，同样一个词或词组，在不同的上下文搭配中，在不同的句法结构中就可能有不同的意义。一个词，一个词组脱离上下文是不能正确理解的。要通读整个句子，重点注意谓语动词、连词、介词、专用词组和实义动词以及暂时不了解的新词汇。决定是否分译以及如何断句，分析原句标点、句型结构、语态及时态特点，注意灵活组织，保持前后文之间的逻辑联系和呼应关系。因此，译者首先应该结合上下文，通过对词义的选择，语法的分析，彻底弄清楚原文的内容和逻辑关系。

（2）表达阶段

表达就是要寻找和选择恰当的归宿语言材料，把已经理解了的原作内容重新叙述出来。在此阶段要决定汉语如何表达、如何组句，所选汉语词意是否确切，以及译文是否要进行查、加、减、改或者引申翻译等，尽量做到概念明确、用词恰当、逻辑清楚、文字通顺。表达的好坏一般取决于理解原著的深度和对归宿语言的掌握程度，理解正确并不意味着表达一定正确。

（3）校对阶段

校对阶段，是理解和表达的进一步深化，是使译文符合标准的一个必不可少的阶段，是对原文内容的进一步核实，对译文语言的进一步推敲。校对对于科技文献的译文来说尤为重要，因为科技文章要求高度精确，公式、数据较多，稍一疏忽就会给工作造成严重的损失。因此，翻译完成后要多读译文，看是否通顺、能懂，上下文及逻辑关系对不对等，既要译者自己懂，也要使别人阅读译文后能理解，要为读者着想。

体会下面句子和短文的翻译。

例：This paper discusses the relation between the sampling period and the stability of sampled-data.

译文：本文讨论了抽样周期与抽样数据稳定性之间的相互关系。

例：The binary system of representing numbers will now be explained by making reference to the familiar decimal system.

译文：下面参照熟悉的十进制系统来说明表示数目的二进制系统。

例：The paper addresses an important problem in database management systems.

译文：本文讨论了数据库管理系统的一个重要问题。

例：The author proposes an approach to the creation of an integrated method of investigation and designing objects, based on a local computer system.

译文：作者创立一种综合法，以本地计算机系统为基础来进行对象研讨与设计。

例：

A computer is an inanimate device that has no intelligence of its own and must be supplied with instructions so that it knows what to do and how and when to do it. These instructions are called software. The importance of software can't be overestimated.

Software is made up of a group of related programs, each of which is a group of related instructions that perform very specific processing tasks. Software acquired to perform a general business function is often referred to as a software package. Software packages, which are usually created by professional software writers, are accompanied by documentation—that explains how to use the software.

这篇短文讲述的是计算机软件，涉及到计算机软件的定义及功能。在对一些词的理解上就要注意准确性及合理性，如 inanimate device、intelligence、instructions、software、programs、software package 及 documentation，应分别翻译成无生命的装置、智能、指令、软件、程序、软件包及文档。

在对这篇短文的关键词理解的基础上，我们可以给出每句话的中文表达。到此，可能还存在对原文表达不准确的地方，因此需要对原文内容进行进一步的核实和推敲。最后，就可以翻译出既忠实于原文，又准确、合理的关于计算机软件这篇短文的译文。

译文：

一台计算机本是一个无生命的装置。它自身并没有智能，必须给它指令，它才能知道去做什么、如何做以及何时去做。这些指令就称为软件。软件的重要性无论如何估计都不会过分。

软件是由一组相互有关的程序组成的，其中的每一个程序都是一组相关指令，而每组指令都将执行极为特定的处理任务。用来完成一般事务性功能任务的软件通常称为软件包。软件包一般均由软件专业人员编写。软件包都带有文档，文档是用来解释如何使用这个软件的。

3．专业英语翻译技巧

（1）原文的分析与理解

要做好翻译工作，必须从深刻理解原文入手，力求做到确切表达原文。原文是翻译的出发点和惟一依据。只有彻底理解原文含义，才有可能完成确切的翻译，才能达到上述翻译标准的要求。要深刻理解原文，首先要认识到专业科技文献所特有的逻辑性、正确性、精密性和专业性等特点，力求从原文所包含的专业技术内容方面去加以理解。其次，要根据原文的句子结构，弄清每句话的语法关系，采用分组归类的方法辨明主语、谓语、宾语及各种修饰语，联系上下文来分析和理解句与句之间、主句与从句之间的关系。专业科技文献中长句难句较多，各种短语和从句相互搭配、相互修饰，使人感到头绪纷繁、无所适从。在这种情况下更应重视语法分析，突出句子骨架，采用分解归类，化繁为简，逐层推进理解的策略。

例：The technical possibilities could well exist, therefore, of nation-wide integrated transmission network of high capacity, controlled by computers, interconnected globally by satellite and submarine cable, providing speedy and reliable communications throughout the world.

这句话看起来挺难理解，但若采用分解归类的语法分析方法则不难对付。首先，能够充当句子谓语的只能是 could well exist，而不可能是其他非限定动词，如 integrated、controlled、interconnected。接下来主语自然是 possibilities，由于 exist 是不及物动词，因此，不存在宾语。进一步分析将发现用作定语的介词短语 of nation-wide ... cable 除了修饰 possibilities 以外没有其他名词可以承受，而分词短语 controlled ...和 interconnected ...又进一步修饰介词短语中的 network。至于 providing ... the world 则显然是表示结果的状态。这样一来，句子的骨架就比较清楚了。本句之所以将定语和被修饰的词分开，是因为定语太长而谓语较短，将谓语提前有助于整个句子结构的平衡。可将该句译为："因此，在技术上完全有可能实现全国性的集成传输网，这种网络容量大，可由计算机控制，并能通过卫星和海底电缆与全球相联系，提供全世界范围内高速、可靠的通信。"

（2）词义的选择与引申

● 词义的选择

在翻译过程中，若英汉双方都是相互对应的单义词时则翻译不成问题，如 ferroalloy（铁合金）。然而，英语词汇来源复杂，一词多义和一词多性的现象十分普遍，比如 power 在数学中译为"乘方"，在光学中译为"率"，在力学中译为"能力"，在电学中译为"电力"。因此，一定要选择合适的词义。

例：The electronic microscope possesses very high resolving power compared with the optical microscope.

译文：与光学显微镜相比，电子显微镜具有极高的分辨率。

例：Energy is the power to do work.

译文：能量是指做功的能力。

● 词义的引申

英汉两种语言在表达方式方法上差异较大，英语一词多义现象使得在汉语中很难找到绝对相同的词。如果仅按词典意义原样照搬，逐字翻译，不仅使译文生硬晦涩，而且会词不达意，造成误解。因此，有必要结合语言环境透过外延看内涵，把词义做一定程度的扩展、引申。

例：Two and three make five.

译文：二加三等于五。（make 本意为"制造"，这里扩展为"等于"）

The report is happily phrased.

译文：报告措词很恰当。（happily 不应译为"幸运地"）

（3）词语的增减与变序

● 词语的增加

由于英语和汉语各自独立演变发展，因而在表达方法和语法结构上有很大的差别。在英译汉时，不可能要求二者在词的数量上绝对相等。通常应该依据句子的意义和结构适当增加、减少或重复一些词，以使译文符合汉语习惯。

例：The more energy we want to send, the higher we have to make the voltage.

译文：想要输电越多，电压也就得越高。（省略 we）

例：This condenser is of higher capacity than is actually needed.

译文：这只电容器的容量比实际所需要的容量大。（补译省略部分的 capacity）

● 词序的变动

英语和汉语的句子顺序通常都是按主语+谓语+宾语排列的，但修饰语的区别却较大。英语中各种短语或定语从句作修饰语时，一般都是后置的，而汉语的修饰语几乎都是前置的，因而在翻译时应改变动词的顺序。同时，还应注意英语几个前置修饰语（通常为形容词、名词和代词）中最靠近被修饰词的为最主要的修饰语，翻译时应首先译出。此外，英语中的提问和强调也大都用倒装词序，翻译时应注意还原。

例：Such is the case.

译文：情况就是这样。（倒装还原）

例：The transformer is a device of very great practical importance which makes use of the principle of mutual induction.

译文：变压器是一种利用互感原理的在实践中很重要的装置。（从句）

（4）词性及成分的转换

● 词性的转换

因为两种语言的词源不同，语法相异，所以会经常遇到两种语言中缺乏词性相同而且词义完全一样的词汇。这时，常视需要改变原文中某些词的词性以适应汉语的表达习惯；实词之间、虚词之间以及实词和虚词之间都可以互换。

例：Extreme care must be taken to the selection of algorithm in program design.

译文：在程序设计中必须注意算法的选择。

上例中将作主语的名称 care 转为动词并作为谓语来翻译，这样更符合汉语表达的习惯。

例：The information here proves invaluable in reaching a conclusion.

译文：这里的资料对于获得结论价值极高。（here 转为形容词）

- 成分的转换

在英译汉时，不能仅依赖语法分析去处理译文，还必须充分考虑汉语的习惯及专业科技文献的逻辑性和严密性。这就要求在翻译过程中，视具体情况将句子的某一成分（主语、谓语、宾语、定语、表语、状语或补语）译成另一成分，或者将短语与短语、主句与从句、短语与从句进行转换。

例：Electronic computers must be programmed before they can work.

译文：必须先为电子计算机编好程序，它才能工作。（从句译为主句）

（5）标点符号的处理

现代英语和汉语在标点符号的使用上大多相同或接近，但是在有些标点符号的使用习惯上，两种语言还是略有差异，如逗号、分号和破折号。在专业英语文献的翻译过程中，对标点符号的适当运用，可以提高译文的可读性。但在表达时应注意按照汉语标点的使用习惯，对原文的某些标点符号进行必要的转换，以避免引起读者理解上的困难或产生歧义。

英语中逗号的使用范围远远大于汉语，其具有汉语的顿号、逗号和其他一些标点符号的作用。在翻译时，可以根据内容和句子结构，将逗号转换为合适的其他标点符号，或者将逗号取消。

英语分号的使用范围较汉语广，也不一定像汉语那样用于连接并列成分。翻译时，应根据情况照搬或转换成其他符号。

英语中的破折号和省略号都较汉语的短，只及汉语中相应符号长度的一半。这两个符号的使用习惯在英语和汉语中大同小异，表达时既可以照搬，也可以转换成其他标点符号。

（6）长句的翻译

使用长句结构，可以说是科技英语的一大特点，计算机专业英语也不例外。在处理长句时，一般采用顺译法、倒译法和分译法。

- 顺译法

当长句的语法结构和时间顺序与汉语基本相同时，可以按照原文顺序直接翻译。

例：Being able to receive information from any one of a large number of separate places, carry out the necessary calculations and give the answer or order to one or more of the same number of places scattered around a plant in a minute or two, or even in a few seconds, computers are ideal for automatic control in process industry.

译文：由于计算机能从工厂大量分散的任何地方获取信息进行必要的运算，并在一两分钟甚至几秒钟内向分散在工厂各处的一处或多处提供响应或发出指令，所以它对加工工业的自动控制是非常理想的。

- 倒译法

当长句的逻辑顺序与汉语相反时，则必须从后面译起，逆着原文的顺序翻译。

例：Aluminum remained unknown until the nineteenth century, because nowhere in nature is it found free, owing to its always being combined with other elements, most commonly with oxygen, for which it has a strong affinity.

译文：铝总是跟其他元素结合在一起；最普遍的是跟氧结合，因为铝跟氧有很强的亲和力。由于这个原因，在自然界找不到游离状态的铝，所以铝直到19世纪才被人发现。

- 分译法

汉语表达的习惯是一个小句子表达一层意思。因此，为了使表达简洁，对于专业英语中的长句，可以将其分解成几个独立的句子，顺序基本不变，并注意前后连贯，同时注意增加一些连词。

例：The structure design itself includes two different tasks, the design of the structure, in which the sizes and locations of the main members are settled, and the analysis of this structure by mathematical or graphical methods or both, to work out how the loads pass through the structure with the particular members chosen.

译文：结构设计包括两项不同的任务：一是结构设计，确定主要构件的尺寸和位置；二是用数学方法或图解方法或二者兼用进行结构分析，以便在构件选定后计算出各载荷通过结构的情况。

（7）翻译科技资料时应注意的问题

首先要把原文全部阅读一遍，了解其内容大意、专业范围和体裁风格，然后开始翻译。如果条件许可，在动手翻译之前能熟悉一下有关的专业知识那就更好了。遇到生词，不要马上查字典，应该先判断是属于普通用语，还是属于专业用语。如果是专业词汇，则要进一步分析是属于哪一个具体学科范围的，然后再有目的地去查找普通词典或有关的专业词典。

翻译时，最好不要看一句译一句，更不能看一个词译一个词。而应该看一小段译一小段。这样做便于从上下文联系中辨别词义，也便于注意句与句之间的衔接以及段与段之间的联系，使译文通顺流畅，而不致成为一句句孤立译文的堆砌。翻译中的汉语表达应避免不符合汉语逻辑以及过于强调"忠实"，而使原文与译文貌合神离。

翻译科技文献并不要求像翻译文艺作品那样在语言形象、修辞手段上花费很大的工夫，但是要求译文必须概念清楚，逻辑正确，数据无误，文字简练，语句流畅。切忌表达啰嗦，中文修辞不当及表达中存在明显的翻译腔。

下面这段文字是关于调制解调器的，我们可以根据前面讲述的翻译方法给出它的译文。

A modem is a device that converts data from digital computer signals to analog signals that can be sent over a phone line. This is called modulation. The analog signals are then converted back into digital data by the receiving modem. This is called demodulation. Modems, in the beginning, were used mainly to communicate between date terminals and a host computer. Later, the use of modems was extended to communicate between end computers. The data rates have increased from 300 bps in early days to 56Kbps today. Nowadays, transmission involves data compression techniques which increase the rates, error detection and error correction for more reliability.

Modem consists of two major components: which are a data pump and a controller. A data pump performs the basic modulation/demodulation tasks, a controller provides the modem's identify: this is where the protocols for hardware error correction and hardware data compression. A traditional modem implements both features in hardware. Some modems use fewer chips compared to traditional modems, which are called soft modems. The work normally done by the missing chips is transferred to software running on the host computer's main processor.

这篇文章共有两段，第一段讲述调制解调器的定义和功能，第二段讲述调制解调器的内部结构。在通读全文的基础上我们必须正确理解以下文中重要的专业词汇或短语，如 modem、digital computer signals、analog signals、modulation、demodulation、data compression techniques、error detection、error correction、data pump、controller、protocols、hardware error correction、hardware data compression 及 soft modems，可分别翻译成：调制解调器、数字计算机信号、模拟信号、调制器、解调器、数据压缩技术、错误检测、错误纠正、数据泵、控制器、协议、硬件纠错、硬件数据压缩及软调制解调器。

译文：

调制解调器是一种将数据从数字信号转换成可以通过电话线传输的模拟信号的器件。上述转换过程叫作调制。在接收端，调制解调器又将模拟信号转换为数字信号，这个过程叫作解调。在一开始，调制解调器主要用来在数据终端和主机间通信，后来，调制解调器的用途扩展到了在终端计算机间通信。数据传输率已从早期的 300bit/s 增加到了如今的 56kbit/s。现在，数据传输采用了提高传输率的数据压缩技术，为了加大可靠性而采用了错误检测和错误纠正技术。

调制解调器有两个主要的组成部分，它们是数据泵和控制器。数据泵完成基本的调制/解调任务，控制器提供调制解调器的硬件纠错、硬件数据压缩功能。传统的调制解调器通过硬件完成上述功能。和传统的调制解调器相比，有些调制解调器使用较少的芯片，这种调制解调器叫作软调制解调器，那些本由芯片完成的工作交给了计算机的中央处理器中运行的软件完成。

1.4.3　专业英语写作

写作是运用语言技巧来表达清楚其完整的思想。专业英语因为其应用目的的不同而体现出其与众不同的特点。由于专业英语文体的主要功能是叙述科技事实、记录科技知识和描述科技新发现，因此，专业英语语言的特征是结构严谨，客观准确。

科技文章属于严肃的书面文体，崇尚严谨周密，要求行文简练，重点突出。行文简练，就是要将论文中可有可无的字、词、句、段落删去，科技文章的分量不在于文章的长短。另外，科技文章中常出现句型、表达方式等一成不变的现象，这种文章读起来晦涩难懂。因此，要尽量做到严谨中见变化，注意句子的长短、结构变化，表达方式的强弱变化和适当运用修辞手法。

要使句子的结构多样化，可以使用并列或主从复合句。在词序上有些句子用主语开头，有些用前置词语开始，有些用子句为先。在科技文章中，为了突出某些材料或观点，常用强调语气，这样可以加强表达力量，从而避免平淡。但要注意强弱相济，不能滥用强调。

在专业英语文章写作中，可以采用以主题句为直线展开的文体或按叙述的内容逐层展开的文体。主题句是能够说明文章或者段落主要思想的句子。根据英语的表达习惯，主题句常常出现在文章或段落的开头，其他内容往往是对它的补充或解释。在行文中，也可以用大量的短句让描述的内容逐层递进，读者应按序阅读，依次领会文章所要描述的知识内容。

例： Decomposition of the ozonide with water afforded levulinic aldehyde, its peroxide, its further oxidation product levulinic acid, and only minite traces of carbon dioxide, formic acid, and succinic acid. （排比句型）

译文： 用水分解臭氧化物给出乙酰丙醛，以及它的过氧化物，进一步的氧化产物乙酰丙酸、微量二氧化碳、甲酸和琥珀酸等。

例：It is often not the computing ability of electronic computer which is required in such case but the ability to store and calculate large number of informations observed.（强调句型）

译文：在这种情况下，通常需要的并非计算机的计算能力，而是存储和处理大量观测信息的能力。

例：In contrast to a stack in which both insertions and deletions are performed at the same end, a queue is a list in which all insertions are performed at one end while all deletions are made at the other. We have already met this structure in relation to waiting lines, where we recognized it as being a first-in, first-out (FIFO) storage system. Actually, the concept of a queue is inherent in any system in which objects are served in the same order in which they arrive.（按主题句直线展开）

译文：栈的插入与删除操作都是在列表的相同端进行的。而与此不同，队列是插入和删除操作分别在两端进行的列表。我们已经遇到过这种与等待队列相关的结构，在此种情况中，我们把它当作是一种先进先出（FIFO）的存储系统。实际上，队列的概念继承于那些对象以到达的顺序接受服务的系统。

例：We can implement a queue in a computer memory within a block of contiguous cells in a way similar to our storage of a stack. Since we need to perform operations at both ends of the structure, we set aside two memory cells to use as pointers instead of just one, as we did for a stack. One of these pointers, called the head pointer keeps track of the head of the queue; the other, called the tail pointer, keeps track of the tail.（按内容逐层递进）

译文：我们可以像存储栈那样通过连续单元组成的存储块在计算机主存储器中实现队列。因为我们需要在此结构的两端都进行操作，因此分配出两个存储单元用来当作指针，而非栈中那样仅仅需要一个单元来存储指针。其中的一个指针被称为头指针，用来保持队列头的轨迹；另一个指针被称为尾指针，用来保持对尾的轨迹。

第 2 章 Hardware Knowledge

2.1 CPU

2.1.1 Text

1. What Is a Processor?

A processor is a functional unit that interprets and carries out instructions. Every processor comes with a unique set of operations such as ADD, STORE, or LOAD that represent the processor's instruction set. Computer designers are fond of calling their computer machines, so the instruction set is sometimes referred to as machine instructions and the binary language in which they are written is called machine language! You should not confuse the processor's instruction set with the instructions found in high-level programming languages, such as BASIC or PASCAL.

An instruction is made up of operations that specify the function to be performed and operands that represent the data to be operated on. For example, if an instruction is to perform the operation of adding two numbers, it must know: (1) What the two numbers are. (2) Where the two numbers are. When the numbers are stored in the computer's memory, they have an address to indicate where they are, so if an operand refers to data in the computer's memory it is called an address. The processor's job is to retrieve instructions and operands from memory and to perform each operation. Having done that, it signals memory to send it the next instruction. This step-by-step operation is repeated over and over again at awesome speed.

The CPU means the Central Processing Unit. It is the heart of a computer system (Fig. 2-1). The CPU in a microcomputer is actually one relatively small integrated circuit or chip. Although most CPU chips are smaller than a lens of a pair of glasses, the electronic components they contain would have filled a room a few decades ago.[1] Using advanced microelectronic techniques, manufacturers can cram tens of thousands of circuits into tiny layered silicon chips that work dependably and use less power.

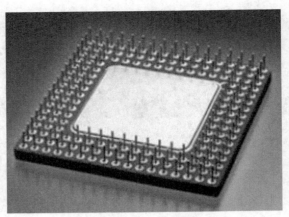

Fig. 2-1 CPU

The CPU coordinates all the activities of the various components of the computer. It determines which operations should be carried out and in what order. The CPU can also retrieve information from memory and can store the results of manipulations back into the memory unit for later reference.

The basic job of computers is the processing of information. For this reason, computers can be defined as devices which accept information in the form of instructions called a program and characters called data, perform mathematical and logical operations on the information, and then supply results of these operations. The program, which tells the computers what to do, and the data, which provide the information needed to solve the problem, are kept inside the computer in a place called memory. [2]

Computers are thought to have many remarkable powers. However, most computers, whether large or small, have three basic capabilities.

First, computers have circuits for performing arithmetic operations, such as: addition, subtraction, multiplication, division and exponentiation.

Second, computers have a means of communicating with the user. After all, if we couldn't feed information in and get results back, these machines would not be of much use.[3]

Third, computers have circuits which can make decisions. The kinds of decisions which computer circuits can make are of the type: Is one number less than another? Are two numbers equal? And, is one number greater than another?

A processor is composed of two functional units: a control unit and an arithmetic / logic unit, and a set of special workspaces called registers.

2. The Control Unit

The control unit is the functional unit that is responsible for supervising the operation of the entire computer system. In some ways, it is analogous to a telephone switchboard with intelligence because it makes the connections between various functional units of the computer system and calls into operation each unit that is required by the program currently in operation.

The control unit is the boss, so to speak, and coordinates all of the CPU's activities. It uses an instruction pointer to keep track of the sequence of instructions that is supposed to be processed.

Using the pointer as a guide, the control unit retrieves each instruction in sequence from RAM and places it in a special instruction register. The control unit interprets the instruction to find out what needs to be done. According to its interpretation, the control unit sends signals to the data bus to fetch data from RAM, and to the ALU to perform a process.

3. The Arithmetic / Logic Unit

The Arithmetic / Logic Unit (ALU) is the functional unit that provides the computer with logical and computational capabilities. Data are brought into the ALU by the control unit, and the ALU performs whatever arithmetic or logic operations are required to help carry out the instructions.

The ALU performs arithmetic computations and logical operations. The arithmetic operations include addition, subtraction, multiplication and division. The logical operations involve comparisons. This is simply asking the computer to determine if two numbers are equal or if one number is greater than or less than another number. These may seem like simple operations. However, by combing these operations, the ALU can execute complex tasks. For example, your video game uses arithmetic operations and comparisons to determine what displays on your screen.

4. Registers

A register is a storage location inside the processor. Registers in the control unit are used to keep track of the overall status of the program that is running. Control unit registers store information such as the current instruction, the location of the next instruction to be executed, and the operands of the instruction. In the ALU, registers store data items that are added, subtracted, multiplied, divided, and compared. Other registers store the results of arithmetic and logic operations.

5. The Fetch and Execute Cycles

At the beginning of each instruction cycle, the CPU fetches an instruction from memory. [4]In a typical CPU, a register called the Program Counter (PC) is used to keep track of the next instruction in sequence. So, for example, consider a computer in which each instruction occupied one 16-bit word of memory, assume that the program counter is set to location 300. The CPU will next fetch the instruction at location 300. On succeeding instruction cycles, it will fetch instruction from locations 301,302,303, and so on.

The fetched instruction is loaded into a register in the CPU known as the instruction register (IR). The instruction is in the form of a binary code that specifies what action the CPU is to take. The CPU interprets the instruction and performs the required action. In general, these actions fall into four categories.

(1) CPU-Memory

Data may be transferred from the CPU to memory or from memory to CPU.

(2) CPU-I/O

Data may be transferred to or from the outside world by transferring between the CPU and an I/O module.

(3) Data Processing

The CPU may perform some arithmetic or logic operation on data.

(4) Control

An instruction may specify that the sequence of execution be altered.[5] For example, the CPU may fetch an instruction from location 149, which specifies that the next instruction be fetched from location 182. The CPU will remember this fact by setting the program counter to 182. Thus, on the next cycle, the instruction will be fetched from location 182 rather than 150. Of course, an instruction's execution may involve a combination of these actions.

6. Performance Factors of CPU

(1) Clock rate

A computer contains a system clock that emits pluses to establish the timing for system operations. The system is not the same as a "real-time clock" that keep track of the time of day. Instead, the system clock sets the speed of "frequency" for data transport and instruction execution. The clock rate set by the system clock determines the speed at which the computer can execute an instruction and, therefore, limits the number of instructions that a computer can complete within a specific amount of time. The time to complete an instruction cycle is measured in megahertz (MHz), or millions of cycles per second.

The microprocessor in the original IBM PC performed at 4.77MHz. Today's processors perform at speeds exceeding 2GHz. If all other specifications are identical, higher megahertz rating mean faster processing.

(2) Word size

Refers to the number of bits that the central processing unit can manipulate at one time. Word size is based on the size of the registers in the CPU and the number of data lines in the bus. For example, a CPU with an 8-bit word size is referred to as an 8-bit processor, it has 8-bit registers and manipulates 8-bit at a time.

A computer with a large word size can process more data in each instruction cycle than a computer with a small word size. Processing more data in each cycle contributes to increased performance. For example, the first microcomputers contained 8-bit microprocessors, but today's faster computers contain 32-bit or 64-bit microprocessor.

(3) Cache

Cache is another factor affecting CPU performance, it is special high speed memory that gives the CPU more rapid access to data. A very fast CPU can execute an instruction so quickly that it often must wait for data to be delivered from RAM, which slow processing. The cache ensures that data is immediately available whenever the CPU requests it.

Key Words

available	可利用的，有效的
analogous	类似的，相似的
comparison	比较，对照
confuse	使混乱
coordinate	（使）协调
establish	建立
exponentiation	幂运算
identical	同一的，完全一样的
megahertz	兆赫兹
occupy	占领，从事
operand	操作数
perform	执行，实现
reference	参考，索引
register	寄存器
remarkable	显著的，不平常的
represent	表示，描述
retrieve	检索，恢复
specification	规格，说明
supervise	监督，指导

Notes

[1] Although most CPU chips are smaller than a lens of a pair of glasses, the electronic components they contain would have filled a room a few decades ago.

译文：虽然大多数 CPU 芯片比一块眼镜片还小，但所包含的电子元件在几十年前却要装满一个房间。

本句由"Although"引导让步状语从句。

[2] The program, which tells the computers what to do and the data, which provide the information needed to solve the problem, are kept inside the computer in a place called memory.

译文：程序的作用是指示计算机如何工作，而数据则是为解决问题提供的所需要的信息，两者都存储在存储器里。

本句的主语是"the program and the data"，由 which 引导的两个定语从句分别修饰 the program 和 the data.

[3] After all, if we couldn't feed information in and get results back, these machines would not be of much use.

译文：如果我们不能输入信息和取出结果，这种计算机毕竟不会有多大用处。

本句的"if we couldn't feed information in and get results back"作条件状语。

[4] At the beginning of each instruction cycle, the CPU fetches an instruction from memory.

译文：在每个指令周期的开始，CPU 都从内存中取一条指令。

本句的主句是"the CPU fetches an instruction"，"At the beginning of each instruction cycle"是时间状语。

[5] An instruction may specify that the sequence of execution be altered.

译文：一条指令可以指定更改执行的顺序。

本句的主语是"An instruction"，"that"引导的是宾语从句，"execution be altered"是定语，修饰"sequence"。

2.1.2　Exercises

1．Translate the following phrases into English or Chinese

（1） machine instructions

（2） binary language

（3） arithmetic computations

（4） current instruction

（5） instruction register

（6） 程序计数器

（7） 系统时钟

（8） 实时时钟

（9） 指令周期

（10） 微电子技术

2．Identify the following to be True or False according to the text

（1） If an operand refers to data in the computer's memory it is called an address.

（2） A register is a storage location outside the processor.

（3） The ALU performs arithmetic computations and logical operations.

（4） A computer contains a system clock that emits pluses to establish the timing for users.

（5） Cache is special high speed memory that gives the CPU more rapid access to data.

（6） The memory ensures that data is immediately available whenever the CPU requests it.

3．Reading Comprehension

（1） The CPU can also retrieve information from memory and can store the results of manipulations back into the ＿＿＿＿＿＿ for later reference.

　　A．ALU

　　B．memory unit

　　C．register

　　D．I/O device

（2） ＿＿＿＿＿＿ is the functional unit that provides the computer with logical and computational capabilities.

　　A．The control unit

　　B．Memory

　　C．The Arithmetic / Logic Unit

　　D．Register

（3） Registers in the ＿＿＿＿＿＿ are used to keep track of the overall status of the program that is running.

　　A．ALU

B. memory
C. I/O system
D. control unit

(4) _____ is used to keep track of which instruction fetch so that it will fetch the next instruction in sequence.
A. Register
B. Controller
C. Storage unit
D. Program counter

(5) The clock rate set by the system clock determines the _____ at which the computer can execute an instruction.
A. speed
B. amount
C. complexity
D. number

2.1.3 Reading Material

Computer Hardware Basics

A computer is a fast and accurate symbol manipulating system that is organized to accept, store, and process data and produce output results under the direction of a stored program of instructions. Fig. 2-2 shows the basic organization of a computer system. Key elements in this system include CPU, the Memory, Input Devices and Output Devices. Let's examine each component of the system in more detail.

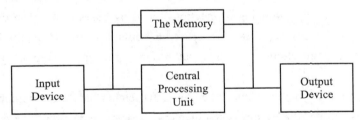

Fig. 2-2 the basic organization of a computer system

1. Central Processing Unit

The processor is the "brains" of the computer that has the ability to carry out our instructions or programs given to the computer. The processor is the part that knows how to add and subtract and to carry out simple logical operations. In a big mainframe computer the processor is called a Central Processing Unit, or CPU, while in a microcomputer, it is usually known as a microprocessor. There are two main sections found in the CPU of a typical personal computer system: the arithmetic-logic section and the control section. But these two sections are not unique to personal computer: They are found in CPUs of all sizes.

2. The Memory

The memory is the computer's work area. There are two types of memory chips: Read Only

Memory (ROM) and Random Access Memory (RAM). ROM chips are designed for applications in which data is only read. (This data can include program instructions) These chips are programmed with data by an external programming unit before they are added to the computer system. Once this is done, the data usually does not change. A ROM chip always retains its data, even when power to the chip is turned off. RAM also called read/write memory, can be used to store data that changes. Unlike ROM, RAM chips lose their data once power is shut off. Many computer systems, including personal computers, include both ROM and RAM.

3. Input Devices

Most of the input devices work in similar ways. The messages or signals received are encoded into patterns which CPU can process by input devices, then, conveyed to CPU. Input devices can not only deliver information to CPU but also activate or deactivate processing just as light switches turn lamps on or off.

4. Output Devices

Output devices can tell the processing results and warn users where their programs or operations are wrong. The most common output devices are monitor, matrix printer, inkjet printer, laser printer, plotter for drawing, speaker, etc. They also work in similar ways. They decode the coded symbols produced by CPU into forms of information that users understand or use easily and show them.

5. System Buses

The components of the computer are connected to the buses. To send information from one component to another, the source component outputs data onto the bus. The destination component then inputs this data from the bus. As the complexity of a computer system increases, it becomes more efficient (in terms of minimizing connections) at using buses rather than direct connections between every pair of devices. Buses use less space on a circuit board and require less power than a large number of direct connections. They also require fewer pins on the chip or chips that comprise the CPU.

Data is transferred via the data bus. When the CPU fetches data from memory, it first outputs the memory address on its address bus. Then memory outputs the data onto the data bus; the CPU can then read the data from the data bus. When writing data to memory, the CPU first outputs the address onto the address bus, then outputs the data onto the data bus. The memory then reads and stores the data at the proper location. The processes for reading data from and writing data to the I/O devices are similar.

The control bus is different from the other two buses. The address bus consists of n lines, which combine to transmit one n-bit address value. Similarly, the lines of the data bus work together to transmit a single multi-bit value. In contrast, the control bus is a collection of individual control signals. These signals indicate whether data is to be read into or written out of the CPU, whether the CPU is accessing memory or an I/O device, and whether the I/O device or memory is ready to transfer data. The control bus is really a collection of (mostly) unidirectional signals. Most of these signals are output from the CPU to the memory and I/O subsystems, although a few are output by these subsystems to the CPU.

2.2 Memory

2.2.1 Text

The memory unit is an essential component in any digital computer since it is needed for storing the programs that are executed by the CPU.[1] People who like PCs always talk about how large their storage device volumes are. But only few really know organization of the storage devices in their PCs.

1. Types of Memory

RAM and ROM play important roles of storage devices. We must know them. RAM is an acronym of Random Access Memory. RAM bars are used for main memory bars. ROM means Read Only Memory. The control and specialized programs are stored in ROM chips and these programs can not be changed by users.

The computer comes with factory-installed permanent memory called ROM. The basic operating instructions are stored in ROM and are not erased when the computer is turned off. In the past, it has been impossible to change the instructions stored in ROM without changing the ROM modules, or the system board in the computer. The computer has a module, called the flash EEPROM (Electrically Erasable Programmable Read-Only Memory) that can be updated. The BIOS (Basic Input/Output System) instructions and the Configuration Utility program are stored in the flash EEPROM in the computer.

In addition to permanent memory, the computer also has a temporary type of memory. The instructions that the computer gets and the information the computer processes remain in RAM during your work sessions. RAM is not a permanent storage place for information. When you turn your computer off, the information you entered during the work session does not remain in memory. Since RAM is only active when the computer is on, the computer uses disk drives to store information even when the computer is off.

Computer memory is measured in kilobytes or megabytes of information. (A byte is the amount of storage needed to hold one character, such as a letter or a numeric digit) One kilobyte(KB) equals 1024 bytes, and one megabyte(MB) is about 1 million bytes. Software requires the correct amount of RAM to work properly. If you want to add new software to your computer, you can usually find the exact memory requirements on the software packaging.

2. The Memory Cells

Memories consist of a number of cells each of which can store a piece of information. Each cell has a number, called its address, by which programs can refer to it. If a memory has n cells, they will have addresses 0 to $n-1$. All cells in a memory contain the same number of bits. If a cell consists of k bits, it can hold any one of 2^k different bit combinations. Note that adjacent cells have consecutive addresses.

Computers that use the binary number system(including octal and hexadecimal notation for binary numbers) also express memory addresses as binary numbers. If an address has m bits, the maximum number of cells directly addressable is 2^m. The number of bits in the address is related to the maximum number of directly addressable cells in the memory and is independent of the number of bits per cell. [2] A memory with 2^{12} cells of 8 bits each and a memory with 2^{12} cells of 64 bits each would each need 12-bit addresses.

The significance of the cell is that it is the smallest addressable unit. In recent years, most computer manufactures have standardized on an 8-bit cell, which is called a byte. Bytes are grouped into words. A computer with a 16-bit word has 2 bytes / word, whereas a computer with a 32-bit word has 4 bytes / word. The significance of a word is that most instructions operate on entire words, for example, adding two words together. Thus a 32-bit machine will have 32-bit registers and instructions for moving, adding, subtracting, and otherwise manipulating 32-bit words. [3]

The 80×86 processors, operating in real mode, have physical address-ability to 1 MB of memory. EMS was developed to allow real mode processing to have access to additional memory. It uses a technique called paging, or bank switching. The requirements for expanded memory include additional hardware and a software device driver. The bank switching registers act as gateways between the physical window within the 1 MB space and the logical memory that resides on the expanded memory board. The device driver, called the Expanded Memory Manager (EMM), controls the registers so that a program's memory accesses can be redirected throughout the entire available expanded memory.

To access expanded memory, a program needs to communicate with the EMM. Communication with the EMM is similar to making calls to DOS. The program sets up the proper CPU registers and makes a software interruption request. More than 30 major functions are defined, and applications and operating systems are given control over expanded memory.

When a program allocates expanded memory pages, the EMM returns a handle to the requesting program. [4] This handle is then used in future calls to the EMM to identify which block of logical pages is being manipulated.

3. Internal Chip Organization

The internal organizations of ROM and RAM chips are similar. To illustrate the simplest organization, a linear organization, consider an 64×4 ROM chip. This chip has six address inputs and four data outputs, and 256 bits of internal storage arranged as sixty-four 4-bit locations.

The six address bits are decoded to select one of the sixty-four locations, but only if the chip enable is active. If CE = 0, the decoder is disabled and no location is selected. The tri-state buffers for that location's cells are enabled, allowing data to pass to the output buffers. If the CE and OE set to 1, these buffers are enabled and the data is output from the chip; otherwise the outputs are tri-stated.

As the number of locations increases, the size of the address decoder needed in a linear organization becomes prohibitively large. To remedy this problem, the memory chip can be designed using multiple dimensions of decoding.

In larger memory chips, this savings can be significant. Consider a 4096×1 chip. The linear organization will require a 12 to 4096 decoder, the size of which is proportional to the number of outputs. (The size of an n to 2^n decoder is thus said to be O (2^n).) If the chip is organized as a 64 × 64 two dimensional array instead, it will have two 6 to 64 decoders: one to select one of the 64 rows and the other to select one of the 64 cells within the row. The size of the decoders is proportional to 2×64, or O ($2 \times 2^{n/2}$) = O($2^{n/2+1}$). For this chip, the two decoders together are about 3 percent of the size of the one larger decoder.

4. Memory Subsystem Configuration

It is very easy to set up a memory system that consists of a single chip.[5] We simply connect the address, data, and control signals from their system buses and the job is done. However, most memory systems require more than one chip. Following are some methods for combining memory chips to form a memory subsystem.

Two or more chips can be combined to create memory with more bits per location (Fig. 2-3). This is done by connecting the corresponding address and control signals of the chips, and connecting their data pins to different bits of the data bus.

Fig. 2-3 Memory

For example, two 64×4 chips can be combined to create an 64×8 memory, as shown in Figure 2-4. Both chips receive the same six address inputs from the bus, as well as the same chip enable and output enable signals. (For now it is only important to know that the signals are the same for both chips). The data pins of the first chip are connected to bits 7 to 4 of the data bus, and those of the other chip are connected to bits 3 to 0.

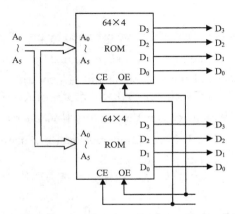

Fig. 2-4 An 64×8 memory subsystem constructed from two 64×4 ROM chips

When the CPU reads data, it places the address on the address bus. Both chips read in address bits A_5 to A_0 and perform their internal decoding. If the CE and OE signals are activated, the chips output their data onto the eight bits of the data bus. Since the address and enable signals are the same for both chips, either both chips or neither chip is active at any given time. The computer never has only one of the two active. For this reason, they act just as a single 64×8 chip, at least as far as the CPU is concerned.

Key Words

acronym	首字母缩略字
activate	激活，使活跃
adjacent	邻近的，接近的
configuration	结构
consecutive	连续的，连贯的
erase	抹去，擦除
gateway	门，通道
kilobyte	千字节
megabyte	兆字节
notation	记号，标记
prohibitively	禁止地
proportional	成比例的
remedy	修补，改善
session	期间
significance	重要性，意义
simplicity	简单，单纯
temporary	临时
turn off	关闭

Notes

[1] The memory unit is an essential component in any digital computer since it is needed for storing the programs that are executed by the CPU.

译文：任何一台数字计算机都需要存储 CPU 所执行的程序，因此，存储器是计算机最重要的部件之一。

本句由"since"引导原因状语从句，"that are executed by the CPU"是定语，修饰"programs"。

[2] The number of bits in the address is related to the maximum number of directly addressable cells in the memory and is independent of the number of bits per cell.

译文：地址的位数与存储器可直接寻址的最大单元数量有关，而与每个单元的位数无关。

本句中，of directly addressable cells in the memory 修饰 the maximum number.

[3] Thus a 32-bit machine will have 32-bit registers and instructions for moving, adding, subtracting, and otherwise manipulating 32-bit words.

译文：因而32位机器有32位的寄存器和指令，以实现传送、加法、减法和其他32位字的操作。

本句的"for moving, adding, subtracting, and otherwise manipulating 32-bit words"是宾语补

足语，进一步解释说明宾语。

[4] When a program allocates expanded memory pages, the EMM returns a handle to the requesting program.

译文：当一个程序装入扩展存储器页中时，EMM就将一个标志回复给这个请求程序。

本句中，由"when"引导了一个时间状语从句。

[5] It is very easy to set up a memory system that consists of a single chip.

译文：构造包含一个简单芯片的存储器系统是非常容易的。

本句中由"It"作形式主语，真正主语是不定式短语，"that consists of a single chip"是定语。

2.2.2 Exercises

1. Translate the following phrases into English or Chinese

（1）storage device volume

（2）RAM bars

（3）permanent memory

（4）electrically erasable programmable read-only memory

（5）expanded memory

（6）软件中断

（7）存储器芯片

（8）存储器子系统

（9）二维阵列

（10）输出缓冲区

2. Identify the following to be True or False according to the text

（1）Most memory systems require only one chip.

（2）Two or more chips can be combined to create memory with more bits per location.

（3）The 80×86 processors, operating in real mode, have physical address-ability to 1 megabyte of memory.

（4）Memories consist of a number of cells（or locations）each of which can store a piece of information.

（5）The internal organizations of ROM and RAM chips are different.

（6）The control and specialized programs are stored in RAM chips.

3. Reading Comprehension

（1）One megabyte equals approximately _____.

A. 1000000 bytes

B. 1024 bytes

C. 65535 bytes

D. 10000 bytes

(2) If a cell consists of n bits, it can hold any one of _____.
A. $2n$ different bit combinations
B. 2^n different bit combinations
C. 2^{n-1} different bit combinations
D. n different bit combinations

(3) When power is removed, information in the semiconductor memory is _____.
A. reliable
B. lost
C. manipulated
D. remain

(4) The six address bits are decoded to select one of the _____ locations, but only if the chip enable is active.
A. thirty-two
B. sixteen
C. six
D. sixty-four

(5) Consider a 4096×1 chip. The linear organization will require a _____ decoder, the size of which is proportional to the number of outputs.
A. 12 to 4096
B. 12 to 244
C. 8 to 256
D. 14 to 4096

2.2.3 Reading Material

Accessing Memory

The instruction cycle is the procedure a microprocessor goes through to process. First the microprocessor fetches, or reads, the instruction from memory. Then it decodes the instruction, determining which instruction it has fetched. Finally, it performs the operations necessary to execute the instruction. Each of these functions—fetch, decode, and execute—consists of a sequence of one or more operations.

Let's start where the computer starts, with the microprocessor fetching the instruction from memory. First, the microprocessor places the address of the instruction on to the address bus. The memory subsystem inputs this address and decodes it to access the sired memory location.

After the microprocessor allows sufficient time for memory to decode the address and access the requested memory location, the microprocessor asserts a READ control signal. The READ signal is a signal on the control bus, which the microprocessor asserts when it is ready to read data from memory or an I/O device. Some processors have a different name for this signal, but all microprocessors have a signal to perform this function. Depending on the microprocessor, the READ signal may be active high (asserted-1) or active low (asserted-0).

When the READ signal is asserted, the memory subsystem places the instruction code to be fetched onto the computer system's data bus, the microprocessor then inputs this data from the bus and stores it in one of its internal registers. At this point, the microprocessor has fetched the instruction.

Next, the microprocessor decodes the instruction. Each instruction may require a different sequence of operations to execute the instruction. When the microprocessor decodes the instruction, it determines which instruction it is in order to select the correct sequence of operations to perform. This is done entirely within the microprocessor; it does not use the system buses.

Finally, the microprocessor executes the instruction. The sequence of operations to execute the instruction varies from instruction to instruction. The execute routine may read data from memory, write data to memory, read data from or write data to an I/O device, perform only operations within the CPU, or perform some combination of these operations.

To read data from memory, the microprocessor performs the same sequence of operations it uses to fetch an instruction from memory.

The symbol, CLK, is the computer system clock. The microprocessor uses the system clock to synchronize its operations. The microprocessor places the address onto the bus at the beginning of a clock cycle. One clock cycle later, to allow time for memory to decode the address and access its data, the microprocessor asserts the READ signal. This causes memory to place its data onto the system data bus. During this clock cycle, the microprocessor reads the data off the system bus and stores it in one of its registers. At the end of the clock cycle it removes the address from the address bus and cancel the READ signal. Memory then removes the data from the data bus, completing the memory read operation.

The processor places the address and data onto the system buses during the first clock cycle of the memory WRITE operation. The microprocessor then asserts a WRITE control signal at the start of the second clock cycle. Just as the READ signal causes memory to read data, the WRITE signal triggers memory to store data. Some time during this cycle, memory writes the data on the data bus to the memory location whose address is on the address bus. At the end of this cycle, the processor completes the memory write operation by removing the address and data from the system buses and canceling the WRITE signal.

One of the most important characteristics of a memory chip is the speed at which data can be accessed from it. To access the data, the address is presented to the address pins, and after a certain amount of time has elapsed, the data shows up at the data pins. The shorter this elapsed time, the better, and consequently, the more expensive the memory chip. The speed of the memory chip is commonly referred to as its access time.

2.3 Input/Output Devices

2.3.1 Text

A computer is a powerful machine that can do everything people assign it. However the computer can't communicate with people directly. By the aid of input and output devices, a computer and people can "know" each other. Using input devices, people "tell" the computer what it should do and the computer feedback the result through output devices.

Input and output devices are the interfaces of man and machine. They usually include keyboard, mouse, monitor, printer, disk (hard disk or floppy disk), input pen, scanner, and microphone, etc. Let's consider some input and output devices that are in common use.

1. Keyboard

The keyboard is used to type information into the computer or input information (Fig. 2-5). There are many different keyboard layouts and sizes with the most common for Latin based languages being the QWERTY layout (named for the first six keys). The standard keyboard has 101 keys. Notebooks have embedded keys accessible by special keys or by pressing key combinations. Some of the keys on a standard keyboard have a special use. There are referred to as command keys. The three most common are the Control or Ctrl, Alternate or Alt and the Shift keys. Each key on a standard keyboard has one or two characters. Press the key to get the lower character and hold Shift to get the upper.

Fig. 2-5 Keyboard

The numeric keypad is located on the right side of the keyboard and looks like an adding machine. However, when you are using it as a calculator, be sure to depress the Num Lock key so the light above Num Lock is lit.

The function keys (F1, F2 and so forth) are usually located at the top of the keyboard. These keys are used to give the computer commands. The function of each key varies with each software program.

The arrow keys allow you to move the position of the cursor on the screen.[1]

Special-purpose keys perform a specialized function. The Esc key's function depends on the program being used. Usually it will back you out of a command. The Print Screen sends a copy of whatever is on the screen to the printer. The Scroll Lock key, which does not operate in all programs, this key is rarely used with today's software. The Num Lock key controls the use of the number keypad. The Caps Lock key controls typing text in all capital letters.

2. Mouse

A mouse is a small device that a computer user pushes across a desk surface in order to point to a place on a display screen and to select one or more actions to take from that position (Fig. 2-6). The most conventional kind of mouse has two buttons on the top: the left one is used most frequently. In the Windows operating systems, it allows the user to click once to send a "Select" indication that provides the user with feedback that a particular position has been selected for further action. The next click on a selected position or quick clicks on it causes a particular action to take place on the selected object. The second button, on the right, usually provides some less-frequently needed capability. For example, when viewing a Web page, you can click on an image to get a pop-up mean that, among other things, you can save the image on your hard disk.

Fig. 2-6 Mouse

A mouse consists of a metal or plastic housing or casing, a ball that sticks out of the bottom of the casing and rolls on a flat surface, one or more buttons on the top of the casing, and a cable that connects the mouse to the computer.[2] As the ball is moved over the surface in any direction, a sensor sends impulses to the computer that causes a mouse-responsive program to re-position a visible indicator (called a cursor) on the display screen. The positioning is relative to some starting place. Viewing the cursor's present position, the user readjusts the position by moving the mouse.

3. Monitor

The monitor shows information on the screen when you type. This is called outputting information. When the computer needs more information, it will display a message on the screen, usually through a dialog box. Monitors come in many types and sizes from the simple monochrome (one color) screen to full color screens (Fig. 2-7).

Fig. 2-7 Monitor

A character-based display divided the screen into a grid of rectangles, each of which can display a single character. The set of characters that the screen can display is not modifiable; therefore, it is not possible to display different sizes or styles of characters. A bitmap display divides the screen into a matrix of tiny, square "dots" called pixels. Any characters or graphics that the computer displays on the screen must be constructed of dot patterns within the screen matrix. The more dots your screen displays in the matrix, the higher its resolution.

- Resolution: Resolution refers to the number of individual dots of color, known as pixels, contained on a display. Resolution is typically expressed by identifying the number of pixels on the horizontal axis (rows) and the number on the vertical axis (columns), such as 640×480. The monitor's viewable area, refresh rate and dot pitch all directly affect the maximum resolution a monitor can display.
- Dot Pitch: Briefly, the dot pitch is the measure of how much space there is between a display's pixels. When considering dot pitch, remember that the smaller the better. Packing the pixels closer together is fundamental to achieving higher resolutions. A display normally can support resolutions that match the physical dot (pixel) size as well as several lesser resolutions.
- Refresh Rate: In monitors based on CRT technology, the refresh rate is the number of times that the image on the display is drawn each second. Refresh rates are very important because they control flicker, and you want the refresh rate as high as possible.

4. Printer

A common type of printer is the matrix printer. A cheap printer might have seven needles, for printing 80 characters in 5*7 matrix across the line. In effect, the print line then consists of 7 horizontal lines, each consisting of 5*80 = 400 dots. Each dot can be printed or not printed, depending on the characters to be printed. The print quality can be increased by two techniques: using more needles and having the circles overlap.

The ink jet printer uses a nozzle and sprays ink onto the paper to form the appropriate characters. In order to get the correct character, the ink is directed with a valve and one or more electronic deflectors that control the vertical and horizontal position of the jet stream of ink. It is possible to print a number of different characters with different styles. Some ink jet printers are capable of printing images in full color.

The laser printer uses laser beams that strike laser-sensitive paper. This paper then picks up a powder or a toner, and the powder or toner is bonded to the paper by heat, pressure, or both. With a laser printer, it is possible to print an entire page at one time. One of the disadvantages of many laser printers is the problem of static electricity. With some of these printers, the paper can have a tendency to stick together. This makes it difficult to output from a laser printer in developing bills and statements sent to customers. A newer cold laser printer has been developed to avoid this problem. They don't require heat in bonding the characters to the paper. As a result, the problems of static electricity and having the paper cling together can be eliminated or substantially reduced.

5. Modem

The need to communicate between distant computers led to the use of the existing phone network for data transmission. Most phone lines were designed to transmit analog information—voices, while the computers and their devices work in digital form—pulses. So, in order to use an analog medium, a converter between the two systems is needed.[3] This converter is the modem. A modem is a device that converts data from digital computer signals to analog signals that can be sent over a phone line.[4] This is called modulation. The analog signals are then converted back into digital data by the receiving modem. This is called demodulation.

Modems can be classified external ones and internal ones. Typically, external modems feature an array of lights set in a display panel that offers important information when you are trying to troubleshoot your setup. You also need a correctly wired cable to connect your modem to an available serial port on your computer. Internal modems are printed circuit boards that take up one of the available expansion slots inside of your computer.

6. Other Devices

With the progress of the technique of input and output devices, scanners gradually come into the ordinary family these years. They can input any kinds of information which are printed on a paper with the most convenient way, by the aid of recognizing and analyzing software, all will be scanned into the computer within several minutes.

If software has music with it, you need a sound box which is attached to your computer to play the music.[5] You may work on the computer while listening to the music. That is really a enjoyment. Have you ever tried to control computer through your voice? Voice control, with the aid of microphone, is already used in some very popular word-processing software.

Key Words

accessible	易接近的，可进入的
bitmap	位图
calculator	计算器
combination	组合，联合
deflector	导向装置
demodulation	解调
embedded	嵌入的，植入的
feedback	反馈
flicker	闪烁，闪动
horizontal axis	水平轴
ink jet printer	喷墨打印机
microphone	麦克风
modifiable	可修正的
modulation	调制
monochrome	单色，黑白
needle	针
nozzle	喷嘴
overlap	重叠，重复
pixel	像素

powder	粉，粉粒
resolution	分辨率
spray	喷射
stick	粘，坚持
tendency	趋势，倾向
troubleshoot	修理故障
vertical axis	垂直轴

Notes

[1] The arrow keys allow you to move the position of the cursor on the screen.

译文：方向键允许你移动光标在屏幕上的位置。

本句中的"to move the position of the cursor on the screen"是宾语补足语。

[2] A mouse consists of a metal or plastic housing or casing, a ball that sticks out of the bottom of the casing and rolls on a flat surface, one or more buttons on the top of the casing, and a cable that connects the mouse to the computer.

译文：鼠标由以下几个部分组成：一个金属或塑料的盒体，一个凸出于盒体底部并可以在平面上滚动的球体，位于盒体上部的一个或多个按键，以及一条连接到计算机的电缆线。

这是一个长句，"consists of"的宾语由若干部分组成，每部分宾语都有自己的定语修饰。

[3] So, in order to use an analog medium, a converter between the two systems is needed.

译文：因此，为了利用传输模拟信号的媒介，在两种系统之间需要一个转换器。

本句中的"So, in order to use an analog medium"是目的状语从句，"between the two systems"是定语，修饰"converter"。

[4] A modem is a device that converts data from digital computer signals to analog signals that can be sent over a phone line.

译文：调制解调器可以将数据从数字信号转换成可以通过电话线传输的模拟信号。

本句中的"that converts data…"是定语从句，修饰"device"，"that can be sent over a phone line"也是定语，修饰"signals"。

[5] If software has music with it, you need a sound box which is attached to your computer to play the music.

译文：如果软件带有音乐，你就需要一个音箱连接到计算机上来播放。

本句由"If"引导条件状语从句。

2.3.2　Exercises

1. Translate the following phrases into English or Chinese

（1）standard keyboard

（2）numeric keypad

（3）function keys

（4）capital letters

（5）pop-up mean

（6）对话框

（7）基于字符的显示器

（8）静电

（9）冷激光打印机

（10）外置式调制解调器

2. Identify the following to be True or False according to the text

（1）Modems can be classified external ones and internal ones.

（2）The most conventional kind of mouse has three buttons on the top.

（3）By the aid of memory, a computer and people can "know" each other.

（4）The monitor shows information on the screen when you type.

（5）Resolution refers to the number of individual dots of color.

（6）The laser printer uses laser beams that strike laser-sensitive paper.

3. Reading Comprehension

（1）The_____are usually located at the top of the keyboard and they are used to give the computer commands.

A. function keys

B. arrow keys

C. numeric keys

D. special-purpose keys

（2）The_____is the measure of how much space there is between a display's pixels.

A. dot pitch

B. scan style

C. solution

D. refresh rate

（3）The_____is the number of times that the image on the display is drawn each second.

A. rate

B. refresh rate

C. solution

D. pixel

（4）The _____uses a nozzle and sprays ink onto the paper to form the appropriate characters.

A. matrix printer

B. laser printer

C. ink jet printer

D. printer

（5）Internal modems are _____that take up one of the available expansion slots inside of your computer.

A. wires

B. buses

C. circuit

D. printed circuit boards

2.3.3 Reading Material
Building a Computer

Building your own computer is not only a valuable skill to possess, but it is also a very simple one to learn and understand.

1. Installing the CPU and RAM

It's much easier to install your CPU while the motherboard is outside of the case, as it gives you much more free space in which to work. First, you will want to make sure that your CPU has either the supplied heat sink or fan attached. Next, simply plug your CPU to the CPU connector. Depending on what type of CPU you are installing, there is only one correct way to install the CPU, gently slide the CPU into position, and check to make sure you have it firmly in place.

You should be able to find your RAM slots located in your main board. Most of today's main boards contain 2 DIMM (Dual in Line Memory Module) connectors. Installing the RAM sticks is a snap-literally. By using DIMM modules it does not matter what size they are, or where/what order you position them in. I find it best to install the first stick in bank 0 (usually the one closest to the CPU) by gently aligning the golden connectors with the slot so that it fits in, apply some pressure, but don't force it in.

Make sure any jumpers are set properly and that your CPU and RAM have already been installed. Now, you will need to plug in the wires that connect the case's LED lights (power, and HD lights), power switch, restart switch and so on. Every main-board needs its case wires arranged differently so consult your manual for correct arrangement of these wires.

2. Installing the IDE devices

Most of today's main boards can support up to four IDE (Integrated Device Electronics) devices. The first thing you will need to do is set your IDE devices' jumper settings. You should find three ways to set the jumpers: master device, slave device, or cable select. It is always a good idea to setup your hard drive (Fig. 2-8) as the master device of your first IDE channel and CD-ROM or DVD-ROM IDE device as the master of the second IDE channel. Remember each IDE channel can only have one master and one slave device.

Fig. 2-8   Hard Disk

3. Throw in the AGP/PCI devices

You simply slide the card into the AGP port (usually the brownish connector closest to the CPU), make sure it is in firmly, and screw the edge connector into the cave. Now you should try booting the computer. If all goes well and the computer POSTs (Power On Self Test), you can power back down and continue to install your PCI cards such as your audio card, modem, and anything else you may need to install. It is good idea to space out your cards so that you can get a good airflow between them. After this is done, get ready for the first real boot.

4. BIOS setup

After your computer success fully POSTs and you have everything installed correctly, you will need to setup the BIOS (Basic Input Output System). Pressing either the Delete or F2 key after the computer POSTs lets you access the computer's BIOS. Flip through your main-board manual for the low-down on accessing and configuring the BIOS.

5. Install the operating system

By far, the easiest operating system to install is Microsoft Windows. Installation is nothing more than booting your computer using either the supplied floppy or CD, and following the simple onscreen prompts during the installation.

When you install the new version of Windows, you can keep an older version of Windows on your computer. More than one operating system installed on a computer is often called a multi-boot configuration. Make sure that your hard disk has a separate partition for each operating system that you want to install, or that your computer has multiple hard disks before you begin the installation. Otherwise, you will either have to reformat and repartition your hard disk or install the new operating system on a separate hard disk. Also, make sure that the partition or disk where you plan to install the new version of Windows is formatted with the NTFS file system. If you have more than one operating system installed on your computer, you can choose which one starts when you turn on your computer.

In fact, software installation is usually easier. Of cause, you should do more practice if you want to master the knowledge of installation better. When using computer, you must know how to communicate with it. In some special situations, the computer will leave some useful error messages or prompts to help you continue with your work. The following are some of the most common messages.

（1）Bad file name or command.

（2）Abort, Retry, Ignore?

（3）Please read the following license agreement.

（4）It is recommended you exit all the other applications before continuing with this installation.

（5）Press OK to continue installation of this software.

（6）Press F1 for Help.

（7）Restart your computer and finish setup.

2.4 专业英语应用模块

2.4.1 The Function of BIOS

BIOS 是计算机的基本输入输出系统，是在出厂前就被主板厂商固化在主板上的一组代码。作为计算机中最基本最重要的设置程序，BIOS 是计算机软件和硬件的纽带，负责计算机的大部分功能，使软件能更好地控制和利用硬件。用户可以通过 BIOS 设置程序来修改硬件参数，从而提高系统的性能。

当计算机启动时，BIOS 首先运行，然后对计算机硬件进行全面检测。在此过程中发现的问题将以两种方式处理：当遇到严重问题时，计算机将停止，不显示任何信号；当遇到其他问题时，计算机将会给出屏幕信息或报警信息并等待用户处理。如果没有遇到问题，计算机硬件将进入准备状态，然后运行系统，并将控制权交给用户。

上面关于 BIOS 的描述信息翻译如下：

BIOS is a set of code solidified in the main board by main board manufacturer before being sold and is the basic input/output system of the computer. As the most fundamental and essential utility of the computer and the tie of software and hardware, BIOS is in charge of the most part of the functions and enable software to control and make use of hardware. User can change some of the parameters of the hardware by the BIOS setup utility to improve the performance of the system.

BIOS boots first when starting the machine, and then BIOS will examine and test the computer hardware completely. The problems found in this process will be dealt in two ways: the machine will stop when meeting serious problems and do not show any signal; the machine will give screen message or sound warning message when meeting other problems and wait for user to do the job. If there is no problem, the hardware will get into ready state, and then start the system and give the control privilege of computer to user.

与 BIOS 的功能描述相关的专业术语如下：

1. Standard CMOS Features   标准 CMOS 功能设定
2. Advanced BIOS Features   高级 BIOS 功能设定
3. Advanced Chipset Features   高级芯片组功能设定
4. Power Management Setup   电源管理设定
5. PnP/PCI Configurations   即插即用/PCI 配置
6. Integrated Peripherals   外部设备设定
7. PC Health status   PC 健康状态
8. Load Optimized Defaults   载入高性能缺省值
9. Set User Password   设置用户密码
10. Set Supervisor Password   设置管理员密码
11. Save & Exit Setup   保存后退出
12. Exit Without Saving   不保存退出

2.4.2 Password Setup

实际上，经常用两种密码来保护计算机。一种用于软件，另一种用于硬件。软件密码又称为 Administrator（管理员），用来保护操作系统，不知道密码的用户不能登录并使用操作系统。

这种密码可以在控制面板中的用户账户中进行设置（见图 2-9）。硬件密码是针对 BIOS 设置程序而言的，称为 Supervisor（管理员）或 User（用户）密码，没有正确的密码不能访问计算机或 BIOS 设置程序。

上面这段文字翻译如下：

In fact, there are two kinds of password usually used to protect the computer, one is to software and the other is to hardware. The software password is the one named Administrator used to protect our Operating System and make user who does not know the password do not log in and use the Operating System. This kind of password can be set with the user account in the Control Panel. The hardware password named supervisor or User is aiming at the BIOS setup utility, the computer or the BIOS setup utility will not be get in without the right password.

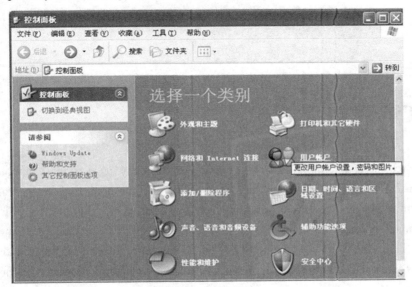

图 2-9 控制面板

计算机密码是让计算机识别用户的一种简便方法。如果密码过于简单，就很容易被破解，即被盗，会造成信息数据的丢失、系统的破坏，后果不堪设想。

设置密码应遵循以下原则：

1. 不要使用和本人有关的姓名，如真实的名字、用户名、注册名等作为密码。
2. 不要使用与本人相关的数字，如出生日期、电话号码、执照号码、门牌号码等信息作为密码。
3. 不要使用任何语言的单个字作为密码。
4. 不用使用"密码（Password）"作为密码。
5. 使用大写字母和小写字母、标点和数字的集合构造密码。
6. 在不同账号里使用不同的密码。
7. 密码至少要 6 个字符，字符数越多，安全性越好。
8. 使用一个方便你记忆的密码，那么就不用写下来了。

第 3 章
Software Knowledge

3.1 Operating System

3.1.1 Text

What exactly does an Operating System (OS) do? Basically, it performs a number of support functions. For example, picture an application program stored on disk. Before the program can be executed, it must first be copied into main memory because the program that controls a computer must be in main memory. The process of copying a program from disk to memory involves considerable logic. The source of a computer's logic is software. Thus, if the application program is to be loaded, there must be a program in memory to control the loading process. That program is the operating system.

Operating systems are either single-tasking or multitasking. The more primitive single-tasking operating systems can run only one process at a time. For instance, when the computer is printing a document, it cannot start another process or respond to new commands until the printing is completed.

All modern operating systems are multitasking and can run several processes simultaneously. In most computers there is only one CPU, so a multitasking operating system creates the illusion of several processes running simultaneously on the CPU. The most common mechanism used to create this illusion is time slice multitasking, whereby each process is run individually for a fixed period of time.[1] If the process is not completed within the allotted time, it is suspended and another process is run. This exchanging of processes is called context switching. The operating system performs the "bookkeeping" that preserves the state of a suspended process. It also has a mechanism, called a scheduler, that determines which process will be run next. The scheduler runs short processes quickly to minimize perceptible delay. The processes appear to run simultaneously because the user's sense of time is much slower than the processing speed of the computer.[2]

Operating systems can use virtual memory to run processes that require more main memory than is actually available.[3] With this technique, space on the hard drive is used to mimic the extra memory needed. Accessing the hard drive is more time-consuming than accessing main memory, however, so performance of the computer slows.

1. Resource Allocation and Related Functions

The resource allocation function allocates resources for use by a user's computation. Resources can be divided into system provided resources like CPUs, memory areas and I/O devices, or user-created resources like files which are entrusted to the OS.

Resource allocation criteria depend on whether a resource is a system resource or a user-created resource. Allocation of system resources is driven by considerations of efficiency of resource utilization. Allocation of user-created resources is based on a set of constraints specified by its creator and typically embodies the notion of access privileges.

Two popular strategies for resource allocation are:
- Partitioning of resources;
- Allocation from a pool.

In the resource partitioning approach, the OS decides a priori what resources should be allocated to a user computation. This approach is called static allocation because the allocation is made before the execution of a program starts. Static resource allocation is simple to implement, however, it could lead to sub-optimal utilization because the allocation is made on the basis of perceived needs of a program, rather than its actual needs. In the latter approach, the OS maintains a common pool of resources and allocates from this pool on a need basis. Thus, OS considers allocation of a resource when a program raises a request for a resource. This approach is called dynamic allocation because the allocation takes place during the execution of a program. Dynamic resource allocation can lead to better utilization of resources because the allocation is made when a program requests a resource.

An OS can use a resource table as the central data structure allocation. The table contains an entry for each resource unit in the system. The entry contains the name or address of the resource unit and its present status, i.e. whether it is free or allocated to some program. When a program raises a request for a resource, the resource would be allocated to it if it is presently free. If many resource units of a resource class exist in the system, a resource request only indicates the resource class and the OS checks if any resource unit of that class is available for allocation.

The CPU can be shared in a sequential manner only. Hence only one program can execute at any time. Other programs in the system have to wait their turn. It is often important to provide fair service to all programs in the system. Hence preemption is used to free the CPU so that it can be given to another program. Deciding which program should be given the CPU and for how long is a critical function. This function is called CPU scheduling, or simply scheduling. Partitioning is a bad approach for CPU sharing, allocation from a pool is the obvious approach to use.

Like the CPU, the memory also can not be shared concurrently. However, unlike the CPU, its availability can be increased by treating different parts of memory as different resources. Both the partitioning and the pool-based allocation approaches can be used to manage the memory resource.[4]

2. Control of I/O Operations

Allocation of a system's resources is closely tied to the operational software's control of I/O operations. As access is often necessary to a particular device before I/O operations may begin, the operating system must coordinate I/O operations and the devices on which they are performed. In effect, it sets up a directory of programs undergoing execution and the devices they must use in completing I/O operations. Using control statements, jobs may call for specific devices. This lets users read data from specific sites or print information at selected offices. Taking advantage of this facility, data read from one location may be distributed throughout computerized system.[5]

To facilitate execution of I/O operations, most operating systems have a standard set of control instructions to handle the processing of all input and output instructions. These standard instructions, referred to as the Input/Output Control System (IOCS), are an integral part of most operating systems. They simplify the means by which all programs being processed may undertake I/O operations.

In effect, the program undergoing execution signals the operating system that an I/O operation is desired, using a specific I/O device. The controlling software calls on the IOCS software to actually complete the I/O operation. Considering the level of I/O activity in most programs, the IOCS instructions are extremely vital.

3. OS Structure

OS design strongly depends on two factors: architectural features of the computer on which it operates, and features of its application domain. Dependence on architectural features is caused by the need to exercise complete control over all functional units of the system. Hence, the OS needs to know the addressing structure, interrupt structure, I/O organization and memory protection features of the computer system. OS policies typically depend on its application domain. For example, the CPU scheduling policy depends on whether the OS will be used for time sharing, or for real time applications. The dependence on these two factors poses obvious difficulties in using an OS on computers with different architectures and different application domains.

Consider the development of an OS for a similar application domain on two computer systems C1 and C2. These two operating systems differ in terms of architecture specific OS code. Remainder of the code, which forms the bulk of the total OS code, does not have any architectural dependencies. It would be tempting to consider development of the OS for C1 and C2 in the following manner.

- Develop the OS for computer system C1. Let this be called OS1.
- Modify OS1 to obtain the OS for computer system C2, i.e. OS2.

That is, OS2 is obtained by porting OS1 to C2.

In early operating systems, this approach faced several difficulties due to the monolithic structure of the OS. Thus, operating systems did not provide clean interfaces between the architecture specific and architecture independent parts of their code. Hence the total porting effort was determined by the total size of OS code, rather than by the size of its architecture specific part. Historically, this difficulty has been addressed by developing an OS structure which separates the architecture specific and architecture size of its architecture specific parts of an OS. This enables operating systems for different computer systems to share much of their design. When operating systems are coded in a high level language, this even permits code sharing across operating systems.

Key Words

allot	分配
concurrently	同时发生地
considerable	相当的，重要的
constraint	强制，约束
criteria	标准
critical	批评的，临界的
domain	领域，范围
embody	使具体化
entrust	信托，委托
illusion	幻影，幻想
integral	整体的
mechanism	机制，机理
mimic	模拟，模仿
monolithic	整体的，完全统一的
partition	分割
perceive	感知，认知
perceptible	可察觉的，感觉得到的
primitive	原始的，基本的
privilege	特权
suspend	暂停，挂起
undergo	经历
utilization	利用

Notes

[1] The most common mechanism used to create this illusion is time slice multitasking, whereby each process is run individually for a fixed period of time.

译文：用于产生这种现象的最常用机制是时间分割多任务处理，以每个过程各自运行固定的一段时间的方式来实现的。

过去分词短语 used to create this illusion 作定语，修饰 mechanism；由 whereby 引导的是非限制性定语从句。whereby：by means of which，以……方式；凭借。

[2] The processes appear to run simultaneously because the user's sense of time is much slower than the processing speed of the computer.

译文：由于用户的时间感觉比计算机的处理速度要慢得多，所以几个程序看起来是同时执行的。

本句中由 "because" 引导原因状语从句，"… much slower than …" 是比较结构。

[3] Operating systems can use virtual memory to run processes that require more main memory than is actually available.

译文：实际可用空间不够时，为了运行那些需要更多主存储空间的程序，操作系统可以利用虚拟存储器。

本句中，to run processes…到句末为目的状语；由 that 引导的定语从句修饰和限定 processes；than 后面省略了主语 that (that 意指 main memory)。

[4] Both the partitioning and the pool-based allocation approaches can be used to manage the memory resource.

译文：资源分区和基于资源池的分配方式都适用于存储器资源管理。

本句的主语是由并列的两部分组成的，即"Both the partitioning and the pool-based allocation approaches"，并且用被动态表示客观性。

[5] Taking advantage of this facility, data read from one location may be distributed throughout computerized system.

译文：利用这一功能，读自某一位置的数据可以分布贯穿整个计算机处理系统。

现在分词短语 Taking…作伴随状语；过去分词短语 read from one location 作定语，修饰主语 data。

3.1.2　Exercises

1．Translate the following phrases into English or Chinese

（1）interrupt structure

（2）architectural feature

（3）context switching

（4）system resource

（5）static allocation

（6）应用程序

（7）动态分配

（8）资源分配

（9）输入/输出控制系统

（10）存储保护属性

2．Identify the following to be True or False according to the text

（1）Operating systems are both single-tasking and multitasking.

（2）The resource allocation function does not allocate resources for use by a user's computation.

（3）With the help of user programs, space on the hard drive is used to mimic the extra memory needed.

（4）All modern operating systems are multitasking and can run several processes simultaneously.

（5）An OS can not use a resource table as the central data structure allocation.

（6）Allocation of a system's resources is closely tied to the operational software's control of I/O operations.

3. Reading Comprehension

(1) _____ criteria depend on whether a resource is a system resource or a user-created resource.
   A. System buses
   B. Resource allocation
   C. Operating system
   D. Register allocation

(2) If many resource units of a resource class exist in the system, a resource request only indicates _____ and the OS checks if any resource unit of that class is available for allocation.
   A. resource allocation
   B. memory class
   C. the resource class
   D. OS class

(3) To facilitate execution of I/O operations, most _____ have a standard set of control instructions to handle the processing of all input and output instructions.
   A. operating systems
   B. software
   C. system data
   D. I/O system

(4) Dependence on _____ is caused by the need to exercise complete control over all functional units of the system.
   A. operation system
   B. software features
   C. architecture
   D. architectural features

(5) When operating systems are coded in a _____, this even permits code sharing across operating systems.
   A. natural language
   B. machine level language
   C. low level language
   D. high level language

3.1.3 Reading Material

Linux Operating System

Linux is an operating system that was initially created as a hobby by a young student, Linus Torvaids, at the University of Helsinki in Finland. He began his work in 1991 when he released version 0.02 and worked steadily until 1994 when version 1.0 of the Linux kernel was released. The kernel, at the heart of all Linux system, is developed and released under the GNU General Public License and its source code is freely available to everyone. It is this kernel that forms the base around which a Linux operating system is developed. There are now hundreds of companies and organizations and an equal number of individuals that have released their own versions of operating systems based on the Linux kernel.

Linux is an implementation of the Unix design philosophy, which means that it is a multi-user system. This has many advantages. Whether one user is running several programs or several users are running one program, Linux is capable of managing the traffic.

Linux is open. This means that all programmers and users can have access to the source code as well as the right to modify it. This means many good things for the user. It means higher-quality software, more efficient and less prone to crash. It also makes it easier for people to get involved in the development process, which means that even if someone is not a programmer, he can have a great impact on a piece of software by suggesting how to improve it to the development team.

Linux is available in several formats. For the technically oriented, there is simply the kernel source code and the various programs and utilities that you can put together. However, for those that do not wish to piece together their own system bit by bit there are various Linux distributions. These distributions are subtly different in the way they set things up and the way they package software. They are similar in that they provide you with some media, usually a set of CDs or floppy disks, from which you can install Linux. They also provide graphical tools to install the system and configure it.

One of the major programs of Linux, however, is installation. It is not as smooth as installing other operating systems. At some point you may well have to delve into the depths of the terse Unix command line. This is partly because Linux is an extremely configurable system and partly a legacy of Unix.

In order to get a better understanding of the way your Linux system works, or how to customize it to suit your needs, you should read the HOWTO documents. These documents explain how one does anything on a Linux system from choosing hardware to setting up. The HOWTOs are available at a lot of Linux resource sites.

Linux is one of the more stable operating systems available today. This is due in large part to the fact that it was written by programmers who were writing for other programmers and not for the corporate system. The only people who make the decisions on what went into the system were programmers. Also, the deadline pressure is not as strong when one is developing as a hobby.

Linux performs well for most applications. However the performance is not optimal under heavy network load. The network performance of Linux is 20%~30% below the capacity of FreeBSD running on the same hardware. The situation has improved somewhat recently and the 2.4 release of the Linux kernel will introduce a new virtual memory system based on the same concepts as the FreeBSD VM system. Since both operating systems are open source, beneficial technologies are shared and for this reason the performance of Linux and FreeBSD is rapidly converging.

Servers often stay up for years. However, disk I/O is non-synchronous by default, which is less reliable for transaction based operations, and can produce a corrupted file system after a system crash or power failure. But for the average user, Linux is a very dependable operating system.

3.2 Data Structures

3.2.1 Text

1. An Introduction to Data Structures

Data comes in all shapes and sizes, but often it can be organized in the same way. For example, consider a list of things to do, a list of ingredients in a recipe, or a reading list for a class, although each contains a different type of data, they all contain data organized in a similar way: a list. A list is one simple example of a data structure. Of course, there are many other common ways to organize data as well. In computing, some of the most common organizations are linked lists, stacks, queues, sets, hash tables, trees, heaps, priority queues, and graphs. Three reasons for using data structures are efficiency, abstraction, and reusability.

Data structures organize data in ways that make algorithms more efficient. For example, consider some of the ways we can organize data for searching it. One simplistic approach is to place the data in an array and search the data by traversing element by element until the desired element is found. However, this method is inefficient because in many cases we end up traversing every element. By using another type of data structure, such as a hash table or a binary tree we can search the data considerably faster.

Data structures provide a more understandable way to look at data; thus, they offer a level of abstraction in solving problems. For example, by storing data in a stack, we can focus on things that we do with stacks, such as pushing and popping elements, rather than details of how to implement each operation. In other words, data structures let us talk about programs in a less programmatic way.

Data structures are reusable because they tend to be modular and context-free. They are modular because each has a prescribed interface through which access to data stored in the data structure is restricted. That is, we access the data using only those operations the interface defines. Data structures are context-free because they can be used with any type of data and in a variety of situations or contexts.[1] In C, we make a data structure store data of any type by using void pointers to the data rather than by maintaining private copies of the data in the data structure itself.

Object-oriented software development is a contemporary approach to the design of reliable and robust software. From the point of view of deciding which data structure should represent that attributes of objects in a specific class, the emphasis that the object—oriented approach places on abstraction is very important to the software development process. Abstraction means hiding unnecessary details.

Procedural abstraction, or algorithmic abstraction, is hiding of algorithmic details, which allows the algorithm to be seen or described, at various levels of detail. Building subprograms so that the names of the subprograms describe what the subprograms do and the code inside subprograms shows how the processes are accomplished is an illustration of abstraction in action. Similarly, data abstraction is the hiding of representational details. An obvious example of this is the building of data types by combining together other data types, each of which describes a piece, or attribute, of a more complex object type. An object-oriented approach to data structures brings together both data abstraction and procedural abstraction through the packaging of the representations of classes of objects.

2. Data Structures and Algorithms

A data structure is a specialized format for organizing and storing data. General data structure types include the array, the file, the record, the table, the tree, and so on.[2] Any data structure is designed to organize data to suit a specific purpose so that it can be accessed and worked with in appropriate ways. In computer programming, a data structure may be selected or designed to store data for the purpose of working on it with various algorithms.

Descriptions of problems in the real world have many superfluous details. An essential step in problem solving is to identify the underlying abstract problem devoid of all unnecessary detail. Similarly, a particular model of computer has many details that are irrelevant to the problem, for example, the processor architecture and the word length. One of the arts of computer programming is to suppress unnecessary detail of the problem and of the computer used.[3]

Once problems are abstracted, it becomes apparent that seemingly different problems are essentially similar or even equivalent in a deep sense. For example, the problems of maintaining a list of students taking a lecture course and of organizing a dictionary structure in a compiler have much in common; both require the storage and manipulation of named things and the things have certain attributes or properties. Abstraction allows common solutions to seemingly different problems. By using abstract algorithms and data structures, a solution to a problem has maximum utility and scope for reuse.

Algorithms and data structures can be specified in any adequately precise language. English and other natural languages are satisfactory if used with care to avoid ambiguity but more precise mathematical languages and programming languages are generally preferred. The execution of the latter can also be automated. A program is the expression of some algorithms and data structures in a particular programming language. The program can be used on all types of computer for which suitable language compilers or interpreters exist, making it more valuable.

Given a problem to solve, we select or design abstract data structures to store the data and algorithms to manipulate them. A data structure is an instance of a data type. The design step is largely independent of any programming language. The coding or programming step is to implement the abstract data structures and algorithms in a particular programming language such as Turing. An abstract data type defines certain static properties of the data and certain dynamic operations on them. It defines what the operations are, not how they work. An algorithm defines some process, how an operation works. It is specified in terms of simple steps. Each step must be clearly implementable in a mechanical way, perhaps by a single machine instruction, by a single programming language statement or by a previously defined algorithm. An algorithm to solve a problem must be correct; it must produce correct output when given valid input data. An algorithm must terminate when given valid input data. When the input data is invalid we usually want the algorithm to warn us and terminate. Often we want to know how much time an algorithm will take or how much space (computer memory) it will use and this leads to the analysis of algorithms and to the field of computational complexity.

3. Data Type

The essence of a data type is that it attempts to identify qualities common to a group of individuals or objects that distinguish it as an identifiable class or kind.[4] If we provide a set of possible data values and a set of operations that act on the values, we can think of the combination as a data type.

Let us look at two classes of data types. We will call any data type whose values we choose to consider atomic an atomic data type. Often we choose to consider integers to be atomic. We are then only concerned with the single quantity that a value represents, not with the fact that an integer is a set of digits in some number system. Integer is a common atomic data type found in most programming languages and in most computer architectures.

We will call any data type whose values are composed of component elements that are related by some structure a structured data type, or data structure. In other words, the values of these data types are decomposable, and we must therefore be aware of their internal construction. There are two essential ingredients to any object that can be decomposed—it must have component elements and it must have structure, the rules for relating or fitting the elements together.

(1) Classes of data types

- Atomic data types (values are not decomposable).
- Data Structures (values are decomposable).

(2) Data Structure—A data type whose values

- Can be decomposed into a set of component data elements each of which is either atomic or another data structure.
- Include a set of associations or relationships (structure) involving the component elements.

A data structure is a data type whose values are composed of component elements that are related by some structure.[5] It has a set of operations on its values. In addition, there may be operations that act on its component elements. Thus we see that a structured data type can have operations defined on its component values, as well as on the component elements of those values.

Key Words

abstraction	抽象，摘要
adequately	足够地，适当地
arithmetic	算术，算法
attribute	属性
considerably	非常地，很
decompose	分解
equivalent	相等的，同等的
essential	重要的，本质的
gracefully	优美地
heap	堆
ingredient	组成，成分
irrelevant	不恰当的，不相干的

modular	有标准组件的
property	性质，属性
recipe	食谱
reusability	可再使用
seemingly	外观上地，表面上地
simplistic	过分简单化的
superfluous	多余的，过剩的
terminate	结束，终止

Notes

[1] Data structures are context-free because they can be used with any type of data and in a variety of situations or contexts.

译文：因为数据结构能用于任何类型的数据，并用于多种环境中，所以数据结构与使用环境无关。

本句中的"because"引导原因状语从句。

[2] General data structure types include the array, the file, the record, the table, the tree, and so on.

译文：一般的数据结构类型包括数组、文件、记录、表、树等。

本句中的"the array, the file, the record, the table, the tree, and so on"是宾语，其中包含若干项并列部分。

[3] One of the arts of computer programming is to suppress unnecessary detail of the problem and of the computer used.

译文：计算机程序设计的艺术之一就是消除一个问题中和所用计算机不必要的细节。

本句中的表语是不定式结构，"of the problem and of the computer used"是定语，修饰"detail"。

[4] The essence of a data type is that it attempts to identify qualities common to a group of individuals or objects that distinguish it as an identifiable class or kind.

译文：数据类型的本质是标识一组个体或目标所共有的特性，这些特性把该组个体作为可识别的种类。

本句由两个复合句构成，均由"that"引导。第一个"that"引导表语从句；第二个"that"引导限定性定语从句，修饰"qualities"，"it"代表"a group of individuals or objects"。

[5] A data structure is a data type whose values are composed of component elements that are related by some structure.

译文：数据结构是一种数据类型，其值是由与某些结构有关的组成元素所构成的。

由"whose"引导的限定性定语从句修饰"a data type"，"that"引导的限定性定语从句修饰"component elements"。

3.2.2 Exercises

1. Translate the following phrases into English or Chinese

（1）object-oriented software

（2）algorithmic abstraction

（3）context-free
（4）stack and queue
（5）natural language
（6）数据类型
（7）抽象数据类型
（8）原子数据类型
（9）结构化数据类型
（10）优先级队列

2. Identify the following to be True or False according to the text
（1）Data structures organize data in ways that make algorithms more efficient.
（2）Data structures provide a more understandable way to look at data.
（3）Abstraction means hiding necessary details.
（4）Algorithms and data structures can not be specified in any adequately precise language.
（5）By using abstract algorithms and data structures, a solution to a problem has maximum utility and scope for reuse.
（6）Any operating system is designed to organize data to suit a specific purpose.

3. Reading Comprehension
（1）Data structures organize data in ways that make＿＿＿＿more efficient.
A. data
B. instruction
C. algorithms
D. memory

（2）＿＿＿＿, we make a data structure store data of any type by using void pointers to the data rather than by maintaining private copies of the data in the data structure itself.
A. Basic language
B. VB language
C. C language
D. Natural language

（3）The program can be used on all types of computer for which suitable ＿＿＿＿ exist, making it more valuable.
A. language compilers or interpreters
B. language compilers
C. language interpreters
D. system software

（4）＿＿＿＿defines certain static properties of the data and certain dynamic operations on them.
A. An abstract data type
B. A data type
C. An abstract class
D. A structure

(5) We will call any data type whose values are composed of component elements that are related by some structure_____.
   A. an abstract data type
   B. a structured data type
   C. a data class
   D. a dynamic data type

3.2.3 Reading Material

Stacks and Queues

A stack is a list in which all insertions and deletions are performed at the same end of the structure. A consequence of this restriction is that the last entry entered will always be the first entry removed—an observation that leads to stacks being known as Last-In, First-Out (LIFO) structures.

The end of a stack at which entries are inserted and deleted is called the top of the stack. The other end is sometimes called the stack's base. To reflect the fact that access to a stack is restricted to the topmost entry, we use special terminology when referring to the insertion and deletion operations. The process of inserting an object on the stack is called a push operation, and the process of deleting an object is called a pop operation. Thus we speak of pushing an entry onto a stack and popping an entry off a stack.

To implement a stack structure in a computer's memory, it is customary to reserve a block of contiguous memory cells large enough to accommodate the stack as it grows and shrinks. Determining the size of this block can often be a critical decision. If too little room is reserved, the stack ultimately exceeds the allotted storage space; if too much room is reserved, memory space will be wasted. One end of this block is designated as the stack's base. It is here that the first entry pushed on the stack is stored, with each additional entry being placed next to its predecessor as the stack grows toward the other end of the reserved block.

Thus, as entries are pushed and popped, the top of the stack moves back and forth within the reserved block of memory cells. A means is therefore needed to maintain a record of the location of the top entry. For this purpose, the address of the top entry is stored in as additional memory cell known as the stack pointer. That is, the stack pointer points to the top of the stack.

Queues occur frequently in everyday life and are therefore familiar to us. The main feature of queues is that they follow a first-come/first-served rule.

There are many applications of the First-In/First-Out (FIFO) protocol of queues in computing. For example, the line of Input/Output (I/O) requests waiting for access to a disk drive in a multi-user time-sharing system might be a queue. The line of computing jobs waiting to be run on a computer system might also be a queue. The jobs and I/O requests are serviced in order of their arrival, that is, the first in is the first out.

A common solution is to set aside a block of memory for the queue, start the queue at one end of the block, and let the queue migrate toward the other end of the block. Then, when the tail of the queue reaches the end of the block, we merely start inserting additional entries back at the original end of the block, which by this time is vacant. Likewise, when the last entry in the block finally becomes the head of the queue and is removed, the head pointer is adjusted back to the beginning of the block where other entries are, by this time, waiting. In this manner, the queue chases itself around within the block rather than wandering off through memory.

Such a technique results in an implementation that is called a circular queue because the effect is that of forming a loop out of the block of memory cells allotted to the queue. As far as the queue is concerned, the last cell in the block is adjacent to the first cell.

3.3 Programming Language

3.3.1 Text

A programming language represents a special vocabulary and a set of grammatical rules for instructing a computer to perform specific tasks. Broadly speaking, it consists of a set of statements or expressions understandable to both people and computers. People understand these instructions because they use human (English and mathematical) expressions. Computers, on the other hand, process these instructions through use of special programs, which, known as translators, decode the instructions from people and create machine language coding.

1. Machine Language

Computer programs that can be run by a computer's operating system are called executables. An executable program is a sequence of extremely simple instructions known as machine code. These instructions are specific to the individual computer's CPU and associated hardware; for example, Intel Pentium and Power PC microprocessor chips each have different machine languages and require different sets of codes to perform the same task. Machine code instructions are few in number (roughly 20 to 200, depending on the computer and the CPU). Typical instructions are for copying data from a memory location or for adding the contents of two memory locations (usually registers in the CPU). Machine code instructions are binary—that is, sequences of bits (0s and 1s). Because these numbers are not understood easily by humans, computer instructions usually are not written in machine code.[1]

2. Assembly Language

Lying between machine languages and high-level languages are assembly languages, which are directly related to a computer's machine language. In other words, it takes one assembly command to generate each machine language command. Machine languages consist entirely of numbers and are almost impossible for humans to read and write. Assembly languages have the same structure and set of commands as machine languages, but they enable a programmer to use names instead of numbers. Thus, assembly language uses commands that are easier for programmers to understand.

Once an assembly-language program is written, it is converted to a machine-language program by another program called an assembler. Assembly language is fast and powerful because of its correspondence with machine language. It is still difficult to use, however, because assembly-language instructions are a series of abstract codes. In addition, different CPU use different machine languages and therefore require different assembly languages. Assembly language is sometimes inserted into a high-level language program to carry out specific hardware tasks or to speed up a high-level program.

3. High-Level Languages

A high-level programming language is a means of writing down, in formal terms, the steps that must be performed to process a given set of data in a uniquely defined way. It may bear no relation to any given computer but does assume that a computer is going to be used.

The high-level languages are often oriented toward a particular class of processing problems. For example, a number of languages have been designed to process problems of a scientific—mathematic nature, and other languages have appeared that emphasize file processing applications.

In a high-level language we would expect to find facilities to do the following:
- structure the program;
- define data elements, give them a name, size and types;
- process data elements (arithmetic / Boolean / transfer);
- control flow of program (test, branch);
- allow commonly used program to be used repeatedly (loops, subroutines, procedures);
- allow input/output of data.

4. C++ and Object-Oriented Programming

C++ is based on C. It is a new version of C. It is a general-purpose and more comprehensive applications programming language developed by Ejarne Stroustrup at Bell Laboratory.[2] C++ retains much of C, including a rich operator set, nearly orthogonal design, terseness, and extensibility. C++ is a highly portable language, and translators for it exist on many different machines and systems. C++ compilers are highly compatible with existing C programs because maintaining such compatibility was a design objective. Unlike other object-oriented languages, such as Smalltalk, C++ is an extension of an existing language widely used on many machines.[3]

C++ is a marriage of the low-level with the high-level. C was designed to be a system implementation language, a language closes to the machine. C++ adds object-oriented features designed to allow a programmer to create or import a library appropriate to the problem domain. The user can write code at the level appropriate to the problem while still maintaining contact with the machine-level implementation details.

C++ fully supports object-oriented programming, including the four pillars of object-oriented development: encapsulation, data hiding, inheritance, and polymorphism.

① Encapsulation and Data hiding

When an engineer needs to add a resistor to the device he is creating, he doesn't typically build a new one from scratch. He walks over to a bin of resistors, examines the colored bands that indicate the properties, and picks the one he needs. The resistor is a "black box" as far as the engineer is concerned—he doesn't much care how it does its work as long as it conforms to his specifications; he doesn't need to look inside the box to use it in his design.

The property of being a self-contained unit is called encapsulation. With encapsulation, we can accomplish data hiding. Data hiding is the highly valued characteristic that an object can be used without the user knowing or caring how it works internally. Just as you can use a refrigerator without knowing how the compressor works, you can use a well-designed object without knowing about its internal data members.

Similarly, when the engineer uses the resistor, he need not know anything about the internal state of the resistor. All the properties of the resistor are encapsulated in the resistor object; they are not spread out through the circuitry. It is not necessary to understand how the resistor works in order to use it effectively. Its data is hidden inside the resistor's casing.

C++ supports the properties of encapsulation and data hiding through the creation of user-defined types, called classes.[4] Once created, a well-defined class acts as a fully encapsulated entity—it is used as a whole unit. The actual inner workings of the class should be hidden. Users of a well-defined class do not need to know how the class works; they just need to know how to use it.

② Inheritance and Reuse

When the engineers at Acme Motors want to build a new car, they have two choices: They can start from scratch, or they can modify an existing model. Perhaps their Star model is nearly perfect, but they would like to add a turbocharger and a six-speed transmission. The chief engineer would prefer not to start from the ground up, but rather to say, "Let's build another Star, but let's add these additional capabilities. We'll call the new model a Quasar." A Quasar is a kind of Star, but one with new features. C++ supports the idea of reuse through inheritance. A new type, which is an extension of an existing type, can be declared. This new subclass is said to derive from the existing type and is sometimes called a derived type.[5] The Quasar is derived from the Star and thus inherits all its qualities, but can add to them as needed.

③ Polymorphism

The new Quasar might respond differently than a Star does when you press down on the accelerator. The Quasar might engage fuel injection and a turbocharger, while the Star would simply let gasoline into its carburetor. A user, however, does not have to know about these differences. He can just "floor it", and the right thing will happen, depending on which car he is driving. C++ supports the idea that different objects do "the right thing" through what is called function polymorphism and class polymorphism. Poly means many, and morph means form. Polymorphism refers to the same name taking many forms.

Key Words

accelerator	加速器
assembler	汇编程序，汇编器
carburetor	化油器，汽化器
encapsulation	封装
executable	可执行的，可实行的
facility	容易，便利，工具
inheritance	继承
injection	注满，充满
morph	变种，形态
orthogonal	正交的，互相垂直的
pillar	台柱，栋梁
polymorphism	多态性，多形性

roughly	粗略地
scratch	凑合的，随意的
subclass	子类
subroutine	子程序
terseness	简洁，简练
turbocharger	涡轮增压器

Notes

[1] Because these numbers are not understood easily by humans, computer instructions usually are not written in machine code.

译文：由于这些数字令人难以理解，所以计算机指令通常不是用机器码来写的。

本句中由"Because"引导原因状语从句，主句用被动语态。

[2] It is a general-purpose and more comprehensive applications programming language developed by Ejarne Stroustrup at Bell Laboratory.

译文：它是由贝尔实验室的 Ejarne Stroustrup 开发的一种通用的、更完整的应用程序设计语言。

本句中的"It"指 C 语言，表语是并列的两部分，"developed by…"是过去分词短语作定语，修饰"language"。

[3] Unlike other object-oriented languages, such as Smalltalk, C++ is an extension of an existing language widely used on many machines.

译文：不像其他的面向对象语言，如 Smalltalk 语言，C++是现有语言的一种扩展，它广泛地用于很多种机器。

本句中的"Unlike other object-oriented languages"是状语，"such as Smalltalk"是同位语。

[4] C++ supports the properties of encapsulation and data hiding through the creation of user-defined types, called classes.

译文：C++通过创建称为类的用户定义类型而支持封装和数据隐藏的属性。

本句中的"through the creation of user-defined types"是状语，而"called classes"是同位语。

[5] This new subclass is said to derive from the existing type and is sometimes called a derived type.

译文：这个新的子类从已存在类派生而来，可称为派生类。

这是一个并列句，主语是"This new subclass"，本句使用被动语态。

3.3.2 Exercises

1．Translate the following phrases into English or Chinese

（1） abstract code

（2） user-defined type

（3） machine language

（4） data hiding

（5） grammatical rule

（6） 汇编语言

（7） 高级语言

（8）面向对象编程
（9）编程语言
（10）机器码

2. Identify the following to be True or False according to the text

（1）Machine languages consist entirely of numbers and are almost impossible for humans to read and write.

（2）Once created, a well-defined class acts as a fully encapsulated entity.

（3）The property of being a self-contained unit is called polymorphism.

（4）Assembly languages have the same structure and set of commands as high-level languages.

（5）C++ is a marriage of the low-level with the high-level.

（6）Different CPU use different machine languages and therefore require different high-level languages.

3. Reading Comprehension

（1）When the engineer uses the resistor, he need know nothing about the _____ of the resistor.

  A. external state
  B. process
  C. internal state
  D. number

（2）_____ may bear no relation to any given computer but does assume that a computer is going to be used.

  A. A high-level programming language
  B. Machine language
  C. A low-level programming language
  D. Assembly language

（3）C++ adds object-oriented features designed to allow a programmer to create or import a library appropriate to _____.

  A. machine code
  B. high-level language
  C. C++ language
  D. the problem domain

（4）An executable program is a sequence of extremely simple _____ known as machine code.

  A. commands
  B. information
  C. instructions
  D. high-level code

(5) _____ might respond differently than a Star does when you press down on the accelerator.
   A. The old Quasar
   B. The new Quasar
   C. A car
   D. Cars

3.3.3 Reading Material

Software Engineering

Software engineering is the application of tools, methods, and disciplines to produce and maintain an automated solution to a real-world problem. It requires the identification of a problem, a computer to execute a software product, and an environment (composed of people, equipment, computers, documentation, and so forth) in which the software product exists.

Software engineering is the discipline of producing software to meet customer needs with the highest quality feasible given resource constraints. It is concerned with the ways in which people conduct their work activities and apply technology to produce and maintain software products and software-intensive systems. Issues of concern include specification, design, implementation, verification, validation, and evolution of software artifacts. Related topics include software metrics, project management, and process improvement. Software engineering is essential for anyone working in development, maintenance, management, or related areas in a software organization.

The overall objective of software engineering is to give the reader a sense of the flow of events in an integrated system and software development effort, and appreciation for and understanding of the software engineer's role in the system development process, and a comprehensive preparation for assuming responsibilities of a software engineer.

Early approaches to software engineering insisted on performing analysis, design, implementation, and testing in a strictly sequential manner. The seeling was that too much was at risk during the development of a large software system to allow for trial-and-error techniques. As a result, software engineers insisted that the entire analysis of the system be completed before beginning the design and, likewise, that the design be completed before beginning implementation. The result was a development process now referred to as the waterfall model, an analogy to the fact that the development process was allowed to flow in only one direction.

In recent years, software engineering techniques have begun to reflect this underlying contradiction as illustrated by the emergence of the incremental model for software development. Following this model, the desired software system is constructed in increments—the first being a simplified version of the final product with limited functionally. Once this version has been tested and perhaps evaluated by the future user, more features are added and tested in an incremental manner until the system is complete. For example, if the system being developed is a student records system for a university register, the first increment may incorporate only the ability to view student records. Once that version is operational, additional features, such as the ability to add and update records, would be added in a stepwise manner.

The incremental model is evidence of the trend in software development toward prototyping in which incomplete versions of the proposed system, called prototypes, are built and evaluated. In the case of the incremental model these prototypes evolve into the complete, final system—a process known as evolutionary prototyping. In other cases, the prototypes may be discarded in favor of a fresh implementation of the final design. This approach is known as throwaway prototyping. An example that normally falls within this throwaway category is rapid prototyping in which a simple example of the proposed system is quickly constructed in the early stages of development. Such a prototype may consist of only a few screen images that give an indication of how the system will interact with the user and what capabilities it will have. The goal is not to produce a working version of the product but to obtain a demonstration tool that can be used to clarify communication between the parties involved. For example, rapid prototypes have proved advantageous in ironing out system requirements during the analysis stage or as aids during sales presentations to potential clients.

3.4 专业英语应用模块

3.4.1 Computer Malfunction

1．引起计算机故障的原因

（1）工作环境引起的故障

静电常造成主板上的芯片被击穿，主板遇到电源损坏或电网电压瞬间产生的尖峰脉冲时，往往会损坏主板供电插头附近的芯片，主板上的灰尘也会造成信号短路。

（2）人为原因引起的故障

带电插拔各种板卡，以及在装板卡及插头时用力不当造成对接口、芯片等的损害。

（3）接触不良引起的故障

各种芯片、插座、接口因锈蚀、氧化等原因而产生的故障。

（4）器件质量问题引起的故障。

（5）软硬件不兼容。

由于软硬件不兼容，造成系统不能启动或不能开机。

（6）短路、断路引起的故障。

2．常见故障

启动计算机时屏幕显示"CMOS Battery Failed"。

故障现象：开机时屏幕显示"CMOS Battery Failed"。

故障原因：表明 CMOS 电池失效或电力不足。

解决方法：更换 CMOS 电池。

上述关于计算机故障分析的信息翻译如下。

1．The reasons causing the malfunctions

（1）About work circumstance

Static electricity often makes the main board punctured, the power damage or the unstable voltage can damage the chip near plug on the main board. And also the dust on the board can make short circuit.

(2) Man-made reasons

Plug cards without cutting the power and damage the interfaces, chips when fabricating plugs and cards with wrong force.

(3) Bad connectivity

The chips, plugs and interfaces can't connect in common for rusting, oxide, and so on.

(4) Devices quality problem

(5) The compatibility between hardware and software

For the compatibility problem, computer can't root up or system can't start.

(6) Short circuit or broken circuit problem

2. Common malfunctions

"CMOS Battery Failed" appears in the screen when you booting the machine.

Phenomenon: This sentence appears in your welcome screen.

Reason: CMOS battery is null or the power is not enough.

Method: Change a new one.

3.4.2 The Malfunctions of Main Board

1. 主板无法启动故障

故障现象：主板无法正常启动，同时发出"滴滴"警报声。

故障原因：出现这种现象的可能原因是，主板内存插槽质量低劣，内存条上的金手指与插槽簧片接触不良；也有可能是内存条上的金手指表面的镀金效果不好，在长时间工作中，镀金表面出现了很厚的氧化层；还有可能是，内存条生产工艺不够标准。

解决方法：将计算机机箱打开，断开电源，取出内存条，将出现在内存条上的灰尘或氧化层用橡皮擦干净，然后重新插入到内存插槽中即可。确保内存条不会左右晃动，这样也能有效避免金手指被氧化。要是上面的方法无法解决故障，可以更换新的内存条试试。

2. 主板散热不良故障

故障现象：计算机频繁死机，在进行 CMOS 设置时也会出现死机现象。

故障原因：这种故障一般是主板散热不良或主板 Cache（缓存）有问题引起的。

解决方法：如果因主板散热不够好而导致该故障，可以在死机后触摸 CPU 周围主板元件，会发现其温度非常高，在更换大功率风扇之后，死机故障即可解决。如果是 Cache 有问题造成的，则可以进入 CMOS 设置，将 Cache 禁止后即可。当然，Cache 禁止后，机器速度肯定会受到影响。如果仍不能解决故障，只有更换主板或 CPU 了。

上述关于主板故障的内容翻译如下。

1. Main board can't start up

Phenomenon: Main board can't start up normally and gives the warning alarm.

Reason: The memory is in a bad quality so that the connectivity between the golden fingers and plug is bad, too; or the golden fingers have the oxide layer with the long time work; or the memory on the main board is not a standard one.

Method: Open your machine, cut the power supply, and then pick up the memory, wipe off the dust with your rubber, and then put memory back to your main board. What you should remember is to make sure your memory is not loose. Or you can change another memory.

2. Bad heat dispersing for main board

Phenomenon: Computer system or CMOS setup utility halted frequently.

Reason: The generally reason is the bad heat dispersing or cache problem.

Method: When the reason is the first one, you can feel the temperature is halted, if the temperature is high, you can change the fan efficiency and then the problem can be solved; if the reason is the second one, you can get into CMOS setup utility and enable the cache. If the reason is not the two given above, you can have a try to change a main board or a CPU.

Database Technology

4.1 Database Principle

4.1.1 Text

1. Fundamental Concepts of Database

A database is a collection of related data. By data, we mean knows facts that can be recorded and that have implicit meaning. For example, consider the names, telephone numbers, and addresses of all the people you know. You may have recorded this data in an indexed address book, or you may have stored it on a diskette using a personal computer and software such as DBASE III or lotus 1-2-3. This is a collection of related data with an implicit meaning and hence is a database.

The above definition of database is general, for example, we may consider the collection of words that make up this page of text to be related data and hence a database. However, the common use of the term database is usually more restricted. A database has the following implicit properties:

- A database is a logically coherent of data with some inherent meaning. A random assortment of data cannot be referred to as a database.
- A database is designed, built, and populated with data for a specific purpose.[1] It has an intended group of users and some preconceived applications in which these users are interested.
- A database represents some aspect of the real world, sometimes called the mini-world. Changes to the mini-world are reflected in the database.

In other words, a database has some source from which data are derived, some degree of interaction with events in the real world, and an audience that is actively interested in the contents of the database.

2. DataBase Management System

These multi-user databases are managed by piece of software called a DataBase Management System (DBMS). It is this which differentiates a database from an ordinary computer file. Between the physical databases itself (i.e. the data as actually stored) and the users of the system is the DBMS. All requests for access to data from users—whether people at terminals or other programs running in batch—are handled by the DBMS.

One general function of the DBMS is the shielding of database users from machine code. In other words, the DBMS provides a view of the data that is elevated above the hardware level, and supports user-requests such as "Get the PATIENT record for patient Smith", written in a higher-level language.

The DBMS also determines the amount and type of information that each user can access from a database. For example, a surgeon and a hospital administrator will require different views of a database.

When a user wishes to access a database, he makes an access request using a particular data manipulation language understood by the DBMS. The DBMS receives the request, and checks it for syntax errors. The DBMS then inspects, in turn, the external schema, the conceptual schema, and the mapping between the conceptual schema and the internal schema. It then performs the necessary operations on the stored data.

In general, fields may be required from several logical tables of data held in the database. Each logical record occurrence may, in turn, require data from more than one physical record held in the actual database.[2] The DBMS must retrieve each of the required physical records and construct the logical view of the data requested by the user. In this way, users are protected from having to know anything about the physical layout of the database, which may be altered, say, for performance reasons, without the users having their logical view of the data structures altered.

A database management system is a collection of programs that enables users to create and maintain a database (Fig. 4-1). The DBMS is hence a general-purpose software system that facilitates the processes of defining, constructing, and manipulating database for various applications. Defining a database involves specifying the types of data to be stored in the database, along with a detailed description of each type of data. Constructing the database is the process of storing the data itself on some storage medium that is controlled by the DBMS. Manipulating a database includes such functions as querying the database to retrieve specific data, updating the database to reflect changes in the mini-world, and generating reports from the data.

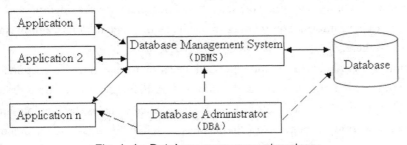

Fig. 4-1 Database management system

3. Logical Structures of DBMS

(1) List structures

In this logical approach, records are linked together by the use of pointers. A pointer is a data item in one record that identifies the storage location of another logically related record. Records in a customer master file, for example, will contain the name and address of each customer, and each record in this file is identified by an account number. During an accounting period, a customer may buy a number of items on different days. Thus, the company may maintain an invoice file to reflect these transactions. A list structure could be used in this situation to show the unpaid invoices at any given time.

(2) Tree structures

In this logical approach, data units are structured in multiple levels that graphically resemble an "upside down" tree with the root at the top and the branches formed below. There's a superior-subordinate relationship in a tree structure. Below the single-root data component are subordinate elements or nodes, each of which, in turn, "own" one or more other elements (or none).

(3) Network structures

Unlike the tree approach, which does not permit the connection of branches, the network structure permits the connection of the nodes in a multidirectional manner. Thus, each node may have several owners and may, in turn, own any number of other data units. Data management software permits the extraction of the needed information from such a structure by beginning with any record in a file.

(4) Relational structures

A relational structure is made up of many tables. The data are stored in the form of "relations" in these tables. For example, relation tables could be established to link a college course with the instructor of the course, and with the location of the class.

4. Information System

The objective of information systems is to provide information to all levels of management at the most relevant time, at an acceptable level of accuracy, and at an economical cost.

Individual businesses require information according to the nature of their operations. A car manufacturer is particularly interested in the extent of competition from overseas manufacturers in the home market and competition from other home-based manufacturers. A tour operator is concerned about purchasing power and its effect on holiday bookings and the political situation prevailing in the various countries. As a general guide, the detail contained in reports containing information varies according to the position of the recipient in the hierarchical management structure.[3] They require information relating to events as they occur so that appropriate action can be taken to control them.[4]

Information systems are often computerized because of the need to respond quickly and flexibly to queries. At the bottom level in the information hierarchy are the transaction processing systems, which capture and process internal information, such as sales, production, and stock data. These produce the working documents of the business, such as invoices and statements. Typically, these are the first systems, which a company will install. Above the transaction-level systems are the decision support systems. These take external information—market trends and other external financial data—and processed internal information, such as sales trends, to produce strategic plans, forecasts, and budgets. Often such systems are put together with PC spreadsheets and other unconnected tools. Management information systems lie at the top of the hierarchy of information needs. The MIS takes the plans and information from the transaction-level systems to monitor the performance of the business as a whole. This provides feedback to aid strategic planning, forecasting, and budgeting, which in turn affects what happens at the transactional level.

The formidable task of the MIS designer is to develop the information flow needed to support decision-making. Generally speaking, much of the information needed by managers who occupy different levels and who have different responsibilities is obtained from a collection of existing information systems (or subsystems).[5] These systems may be tied together very closely in a MIS. More often, however, they are more loosely coupled.

Key Words

appropriate	适当的
assortment	分类，各种各样
audience	观众
coherent	一致的，连贯的
database	数据库
differentiate	区别，区分
elevate	举起，提拔
extraction	抽出，取出
formidable	艰难的，令人敬畏的
implicit	暗示的，固有的
interaction	交互作用
invoice	发票
preconceive	事先认为
prevailing	优势的，主要的
purchase	购买
reflect	反映
resemble	相似，相类
restrict	限制，限定
schema	模式
shielding	屏蔽，防护层
surgeon	外科医生
transaction	事务

Notes

[1] A database is designed, built, and populated with data for a specific purpose.

译文：数据库是由用于某种特定目的的数据设计、构造和提供的。

本句的谓语动词有三个，即"designed, built, and populated"，用被动态表示客观性，"for a specific purpose"是宾语补足语。

[2] Each logical record occurrence may, in turn, require data from more than one physical record held in the actual database.

译文：每个逻辑记录值可能依次需要多个保存在实际数据库物理记录中的数据。

本句的"in turn"是插入语，"from more than…"是宾语补足语，进一步说明宾语"data"。

[3] As a general guide, the detail contained in reports containing information varies according to the position of the recipient in the hierarchical management structure.

译文：作为一般性的指导，信息报告的细节随接受者在管理阶层中的位置而变化。

本句中，过去分词短语"contained in reports"作定语，修饰"detail"；现在分词短语"containing information"同样作定语，修饰"reports"；"according to…"作伴随状语。

[4] They require information relating to events as they occur so that appropriate action can be taken to control them.

译文：他们需要与事件发生时相关的信息以便采取适当举措加以控制。

本句中，分词短语"relating to events"作定语，修饰"information"；"as they occur"为伴随状语；"so that…"为目的状语从句。

[5] Generally speaking, much of the information needed by managers who occupy different levels and who have different responsibilities is obtained from a collection of existing information systems (or subsystems).

译文：一般而言，不同级别、不同职能的管理者所需的信息大多来自现有的信息系统（或子系统）集。

本句的"needed by managers"作定语，修饰"information"；"who…and who…"为两个并列的定语从句，均修饰"managers"。

4.1.2　Exercises

1. Translate the following phrases into English or Chinese

（1）Database management system

（2）Management information system

（3）user-request

（4）syntax error

（5）tree structure

（6）多用户数据库

（7）数据处理语言

（8）概念模式

（9）关系结构

（10）决策支持系统

2. Identify the following to be True or False according to the text

（1）A database has some source from which data are derived.

（2）A database is a collection of related data.

（3）The network structure permits the connection of the nodes in a multidirectional manner.

(4) In general, fields may be required from several logical tables of data held in the database.
(5) A network structure is made up of many tables.
(6) There's a superior-subordinate relationship in a network structure.

3. Reading Comprehension

(1) A database represents some aspect of _____, sometimes called the mini-world.
A. the logical world
B. the real world
C. information
D. data

(2) The _____ receives the request, and checks it for syntax errors.
A. MIS
B. DBMS
C. database
D. administrator

(3) The _____ must retrieve each of the required physical records and construct the logical view of the data requested by the user.
A. data processor
B. database
C. DBMS
D. MIS

(4) Manipulating a database includes such functions as _____.
A. querying the database to retrieve specific data
B. updating the database to reflect changes in the mini-world
C. generating reports from the data
D. All above are right

(5) The _____ takes the plans and information from the transaction-level systems to monitor the performance of the business as a whole.
A. MIS
B. DBMS
C. database
D. administrator

4.1.3 Reading Material

Introduction to Typical Databases

1. SQL Server

Microsoft SQL Server is the first scaleable high-performance database management system designed to meet the demanding requirements of distributed client/server computing. Microsoft SQL Server provides the following function.

- Integration with Microsoft Windows NT threading and scheduling services, Performance Monitor, and Event Viewer. A single Windows NT logon to both the network and SQL Server simplifies management of user accounts.
- Built-in replication for reliable dissemination of information throughout an enterprise, reducing the risk of downtime, and putting timely, accurate information close to people who need it.
- Parallel architecture. By executing internal database functions in parallel, the performance and scalability of the system is dramatically increased.
- Centralized management of servers throughout the enterprise with the comprehensive distributed framework. A Windows-based management interface provides visual drag-and-drop control over multiple servers for remote management of data replication, server administration, diagnostics, and tuning.
- Better support for very large databases by taking advantage of parallel architecture. Reduces I/O for many development and maintenance tasks.
- A library of OLE Distributed Management Objects that are available in the Distributed Management Framework.

2. Oracle

As part of every database server, Oracle provides the Oracle Enterprise Manager (EM), a database management tool framework with a graphical interface used to manage database users, instances, and features that can provide additional information about the Oracle environment.

In the EM repository for Oracle9i, the super administrator can define services that should be displayed on other administrations' consoles, and management regions can be set up.

Oracle Enterprise Manager can be used for managing Oracle Standard Edition or Enterprise Edition. Additional functionality for diagnostics, tuning, and change management of Standard Edition instances is provided by the Standard Management Pack. For Enterprise Edition, such additional functionality is provided by separate Diagnostics, Tuning, and Change Management Packs.

As every database administrator knows, backing up a database is a rather mundane but necessary task. Typical backups include complete database backups, table-space backups, data-file backups, control file backups, and archive log backups. Oracle8i introduced the Recovery Manager (RMAN) for the service-managed backup and recovery of the database. Previously, Oracle's Enterprise Backup Utility (EBU) provided a solution on some platforms. However, RMAN, with its Recovery Catalog stored in an Oracle database, provides a much more complete solution. RMAN can automatically locate, back up, restore, and recover data-files, control files, and archived redo logs. RMAN for Oracle9i can restart backups and restores and implement recovery window policies when backups expire. The Oracle Enterprise Manager Backup Manager provides a GUI-based interface to RMAN.

3. Database Technology on the Web

The Web provides access to a variety of data that can be multimedia-rich. Readily available information retrieval techniques such as inverted indices, which allow efficient keyword-based access to text, largely enabled access to the exponentially growing Web. As pressure from users mounted to allow access to richer types of information and to provide services beyond simple keyword-based search, the database research community with a two-pronged solution. First, by using databases to model Web pages, information could be extracted to dynamically build a schema against which users could submit SQL-like queries. By adopting XML for data representation, the second proposed solution centered on adding database constructs to HTML to provide richer, queriable data types.

Today's DBMS technology faces yet another challenge as researchers attempt to make sense of the immense amount of heterogeneous, fast-evolving data available on the Web. The large number of cooperating databases greatly complicates autonomy and heterogeneity issues and requires a careful scalable approach. We need better models and tools for describing data semantics and specifying metadata.

4.2 Data Warehouse and Data Mining

4.2.1 Text

Advances in computer and networking technology have led to the introduction of very powerful hardware and software platforms that can collect, manage, and distribute large amounts of pertinent data.[1] In the case of a business application, detailed transactions are often generated during product-or service-related interactions. These transactions are not limited to commercial sectors. They are also found in sectors such as, government, health care, insurance, manufacturing, finance, distribution, education, and so on. Any enterprise that has some computerized record keeping systems and is interested in deducting or drawing logical conclusions from their voluminous, granular, and detailed information pool should consider building an enterprise-level data warehouse application. These enterprises will then be capable of improving their insights into the trends in their operations and eventually increase the accuracy of their forecasts and plans. The effectiveness of the data warehouse application intensifies especially when the operational data resides in distributed, non-homogenous systems and replace manual data gathering and reconciliation procedures.

1. Data warehouse

Many companies have allowed their data to be stored in many separate systems that are unable to provide a consolidated view of information usable company-wide. One way to address this problem is to build a data warehouse. A data warehouse is a database that consolidates data extracted from various production and operational systems into one large database that can be used for management reporting and analysis.[2] The data from the organization's core transaction processing systems are reorganized and combined with other information, including historical data so that they can be used for management decision making and analysis (Fig. 4-2).

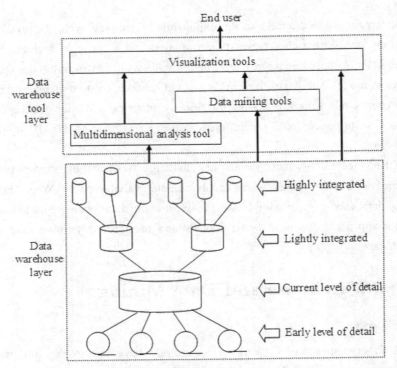

Fig. 4-2   Data warehouse system structure

    A data warehouse is becoming more of a necessity than an accessory for a progressive, competitive, and focused organization. It provides the right foundation for building decision support and executive information system tools that are often built to measure and provide a feet for how well an organization is progressing toward its goals.

    Operation data is the highly structured sets of information that support the ongoing and day-to-day operation of an organization. In case of a decentralized organization, operational data is generated at remote location sometimes in non-homogeneous distributed systems. Distributed systems can span many different geographical locations and time zones. They are configured to provide scalability, visibility, and tracking capabilities of business processes.

    The infrastructure that supports the data warehouse application relies on the same technologies that most other applications are dependent upon. The difference is in the variety and specialization at the product level that can greatly improve the quality of the data warehouse infrastructure.

    Below are some technologies that have made their mark in the data warehouse marketplace. In order to produce a data warehouse that best meets user's needs, these underlining technologies have to be evaluated as part of the periodic resource capacity planning. Depending upon the requirements and resources available, the best combination can be selected and configured.

- Server technology.
- Client technology.
- Database Management System (DBMS) technology.
- Networking technology.
- Mass storage technology.

- Data presentation and publication requirements.
- Software engineering methodology and tools.

In most cases, the data in the data warehouse can be used for reporting. They cannot be updated so that the performance of the company's underlying operational system is not affected. The focus on problem solving describes some of the benefits companies have obtained by using data warehouses.

Data warehouses often contain capabilities to remodel the data. A relational database allows views of data into two dimensions. A multidimensional view of data lets users look at data in more than two dimensions. For example, sales by region by quarter. To provide this type of information, organizations can either use a specialized multidimensional database or tool that takes multidimensional views of data in relational databases.[3] Multidimensional analysis enables users to view the same data in different ways using multiple dimensions. Each aspect of information—product, pricing, cost, region, or time period—represents a different dimension. So a product manager could use a multidimensional tool to learn how many items were sold in southwest sales region in July, how that compares with the previous month and the previous July, and how it compares with the sales forecast.

Among the issues to be addressed in building a warehouse are the following:
- When and how to gather data. In a source-driven architecture for fathering data, the data sources transmit new information, either continually, as transaction processing takes place, or periodically, such as each night. In a destination-driven architecture, the data warehouse periodically sends requests for new data to the sources.

Unless updates at the sources are replicated at the warehouse via two-phase commit, the warehouse will never be quite up to date with the sources.[4] Two-phase commit is usually far two expensive to be an option, so data warehouses typically have slightly out-of-date data.[5] That, however, is usually not a problem for decision-support systems.
- What schema to use. Data sources that have been constructed independently are likely to have different schemas. In fact, they may even use different data models. Part of the task of a warehouse is to perform schema integration, and to convert data to the integrated schema before they are stored. As a result, the data stored in the warehouse are not just a copy of the data at the sources. Instead, they can be thought of as a stored view (or materialized view) of the data at the sources.
- How to propagate updates. Updates on relations at the data sources must be propagated to the data warehouse. If the relations at the data warehouse are exactly the same as those at the data source, the propagation is straightforward.
- What data to summarize. The raw data generated by a transaction-processing system may be too large to store on-line. However, we can answer many queries by maintaining just summary data obtained by aggregation on a relation, rather than maintaining the entire relation. For example, instead of storing data about every sale of clothing, we can store total sales of clothing by category.

2. Data mining

Data mining is about analyzing data and finding hidden patterns using automatic or semiautomatic means. Data mining provides a lot of business value for enterprises.

- Increasing competition
- Customer segmentation
- Churn analysis
- Cross-selling
- Sales forecast
- Fraud detection
- Risk management

During the past decade, large volumes of data have been accumulated and stored in databases. Much of this data comes from business software, such as financial applications, Enterprise Resource Planning (ERP), Customer Relationship Management (CRM), and Web logs. The result of this data collection is that organizations have become data-rich and knowledge-poor. The collections of data have become so vast and are increasing so rapidly in size that the practical use of these stores of data has become limited. The main purpose of data mining is to extract patterns from the data at hand, increase its intrinsic value and transfer the data to knowledge.

Data mining applies algorithms, such as decision trees, clustering, association, time series, and so on, to a dataset and analyzes its contents (Fig. 4-3). This analysis produces patterns, which can be explored for valuable information. Depending on the underlying algorithm, these patterns can be in the form of trees, rules, clusters, or simply a set of mathematical formulas. The information found in the patterns can be used for reporting, as a guide to marketing strategies, and, most importantly, for prediction.

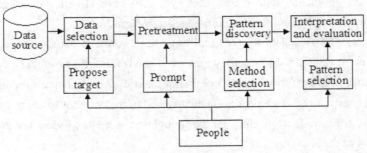

Fig. 4-3   Data mining

Key Words

churn	搅拌，制造
fraud	欺骗，冒牌货
granular	颗粒状的
insurance	保险，预防措施
interaction	互动，交互
multidimensional	多维的
out-of-date	过时的，落伍的

periodically	周期性地，定期地
pertinent	相关的，关于……的
prediction	预报，预言
segmentation	分割，分段，切分
voluminous	大的，容量大的

Notes

[1] Advances in computer and networking technology have led to the introduction of very powerful hardware and software platforms that can collect, manage, and distribute large amounts of pertinent data.

译文：计算机和网络技术的进步导致出现了功能非常强大的，能收集、管理和分发大量有关数据的硬件和软件平台。

本句的"that can collect…"作定语，修饰"hardware and software platforms"。

[2] A data warehouse is a database that consolidates data extracted from various production and operational systems into one large database that can be used for management reporting and analysis.

译文：数据仓库是一种数据库，它将从不同产品和操作系统调出的数据组合起来放入这种大型数据库，对管理状况做出报告和进行分析。

"that consolidates…"作定语，修饰"database"，"that can be used…"作定语，修饰"one large database"。

[3] To provide this type of information, organizations can either use a specialized multidimensional database or tool that takes multidimensional views of data in relational databases.

译文：为了提供这种信息，组织可以用一种特殊化的多维数据库，或用可以在关系数据库中生成数据的多维视图的工具。

本句的"To provide this type of information"是目的状语，"that takes multidimensional…"作定语，修饰"tool"。

[4] Unless updates at the sources are replicated at the warehouse via two-phase commit, the warehouse will never be quite up to date with the sources.

译文：除非对源的更新通过两阶段提交在数据仓库中做了复制，否则数据仓库不可能总是与源同步。

本句中"unless updates at…"作条件状语从句。

[5] Two-phase commit is usually far two expensive to be an option, so data warehouses typically have slightly out-of-date data.

译文：两阶段提交通常因开销太大而不被采用，所以数据仓库常会保留稍微有点儿过时的数据。

本句中的"so data warehouse…"作结果状语从句。

4.2.2　Exercises

1. Translate the following phrases into English or Chinese

（1）marketing strategy

（2）sales forecast

（3）enterprise-level data warehouse application
（4）data presentation
（5）risk management
（6）决策树
（7）数据仓库
（8）操作数据
（9）海量存储技术
（10）数据挖掘

2. Identify the following to be True or False according to the text

（1）A data warehouse is becoming more of a necessity than an accessory for a progressive, competitive, and focused organization.

（2）The data from the organization's core transaction processing systems are reorganized and combined with other information.

（3）The infrastructure that supports the data warehouse application does not rely on the same technologies that most other applications are dependent upon.

（4）Updates on relations at the data sources must not be propagated to the data warehouse.

（5）Data mining is about analyzing data and finding hidden patterns using automatic or semiautomatic means.

（6）Operation data isn't the highly structured sets of information that support the ongoing and day-to-day operation of an organization.

3. Reading Comprehension

（1）Data mining applies algorithms, such as decision trees, clustering, association, time series, and so on, to a dataset and analyzes its _____.
   A. information
   B. contents
   C. data
   D. database

（2）Many companies have allowed their data to be stored in many _____ that are unable to provide a consolidated view of information usable company-wide.
   A. private systems
   B. public systems
   C. hybrid systems
   D. separate systems

（3）In case of a decentralized organization, operational data is generated at remote location sometimes in _____ systems.
   A. homogeneous distributed
   B. non-homogeneous
   C. non-homogeneous distributed
   D. homogeneous

(4) If the relations at the data warehouse are exactly the same as those at the _____, the propagation is straightforward.
  A. data source
  B. server
  C. storage
  D. cloud computing

(5) The raw data generated by a transaction-processing system may be too _____ to store on-line.
  A. large
  B. small
  C. difficult
  D. complex

4.2.3 Reading Material

Expert System

The reliance on the knowledge of a human domain expert for the system's problem solving strategies is a major feature of expert systems. An expert system is a set of programs that manipulate encoded knowledge to solve problems in a specialized domain that normally requires human expertise.

The expert knowledge must be obtained from specialists or other sources of expertise, such as texts, journal articles, and data bases. This type of knowledge usually requires much training and experience in some specialized field such as medicine, geology, system configuration, or engineering design. Once a sufficient body of expert knowledge has been acquired, it must be encoded in some form, loaded into a knowledge base, then tested, and refined continually throughout the life of the system.

Expert systems differ from conventional computer systems in several important ways.

1. Expert systems use knowledge rather than data to control the solution process. Much of the knowledge used is heuristic in nature rather than algorithmic.

2. The knowledge is encoded and maintained as an entity separate from the control program. As such, it is not compiled together with the control program itself. This permits the incremental addition and modification of the knowledge base without recompilation of the control programs.

3. Expert systems are capable of explaining how a particular conclusion was reached, and why requested information is needed during a consultation.

4. Expert systems use symbolic representations for knowledge and perform their inference through symbolic computations that closely resemble manipulations of natural language.

5. Expert systems often reason with meta-knowledge; that is, they reason with knowledge about themselves, and their own knowledge limits and capabilities.

The reasoning of an expert system should be open to inspection, providing information about the state of its problem solving and explanations of the choices and decisions that the program is making. Explanations are important for a human expert, such as a doctor or an engineer, if he or she is to accept the recommendations from a computer.

The exploratory nature of AI and expert system programming requires that programs be easily prototyped, tested, and changed. AI programming languages and environments are designed to support this iterative development methodology. In a pure production system, for example, the modification of a single rule has no global syntactic side effects. Rules may be added or removed without requiring further changes to the large program. Expert system designers often comment that easy modification of the knowledge base is a major factor in producing a successful program.

A further feature of expert systems is their use of heuristic problem-solving methods. As expert system designers have discovered, informal "tricks of the trade" and "rules of thumb" are an essential complement to the standard theory presented in textbooks and classes. Sometimes these rules augment theoretical knowledge in understandable ways; often they are simply shortcuts that have, empirically, been shown to work.

It is interesting to note that most expert systems have been written for relatively specialized, expert level domains. These domains are generally well studied and have clearly defined problem-solving strategies. Problems that depend on a more loosely defined notion of "common sense" are much more difficult to solve by these means. Current deficiencies include:

- Difficulty in capturing "deep" knowledge of the problem domain.
- Lack of robustness and flexibility.
- Inability to provide deep explanations.
- Difficulties in verification.
- Little learning from experience.

In spite of these limitations, expert systems have proved their value in a number of important applications.

4.3 Big Data and Cloud Computing

4.3.1 Text

1. Big Data

Big data usually refers to data in the petabyte and exabyte range—in other words, billions to trillions of records, all from different sources. Big data are produced in much larger quantities and much more rapidly than traditional data (Fig. 4-4). Even though "tweets" are limited to 140 characters each, Twitter generates more than 8 terabytes of data daily. According to the IDC technology research firm, data is more than doubling every two years, so the amount of data available to organizations is skyrocketing.[1] Making sense out of it quickly in order to gain a market advantage is critical.

Fig. 4-4　Big data

Businesses are interested in big data because they contain more patterns and interesting anomalies than smaller data sets, with the potential to provide new insight into customer behavior, weather patterns, financial market activity, or other phenomena. However, to derive business value from these data, organizations need new technologies and tools capable of managing and analyzing nontraditional data along with their traditional enterprise data.

The enormous growth of data has coincided with the development of the "cloud", which is really just a new alias for the global Internet. As data proliferates, especially with the growth of mobile data applications using Apple and Android smart phones and operating systems, the requirements for data transmission and storage also grow. No single site server configuration offers sufficient capacity for the enormous amounts of data derived from sites such as Microsoft.com or FedEx.com, which serve global communities of customers, with transactions numbering in the millions.[2] Users want fast response times and immediate results, especially in the largest and most complex web-based business environments, operating at Internet speed.

Big data analysis requires a cloud-based solution—a networked approach capable of keeping up with the accelerating volume and speed of data transmission. This means that clusters of servers and software such as Hadoop and enterprise control language (ECL), which help find hidden meaning in large amounts of data, are distributed throughout the cloud on high-bandwidth cables. Using the cloud-based approach, a Google marketing analyst may be sitting in Silicon Valley inputting information requests, but the data is being processed simultaneously in Tokyo, Amsterdam and Austin, Texas.

The largest companies making the most sophisticated industrial equipment are now able to go from merely detecting equipment failures to predicting them. This allows the equipment to be replaced before a serious problem develops.

Large Internet commerce companies such as Amazon and Taobao use big data analytics to predict buyer activity as well as to understand warehousing requirements and geographic positions.[3]

Government medical agencies and medical scientists use big data for early discovery and tracking of potential epidemics. A sudden increase in emergency room visits, or even increased sales of certain over-the-counter drugs, can be early warnings of communicative disease, allowing doctors and emergency response officials to activate control and containment procedures.

Many big data applications were created to help Web companies struggling with unexpected volumes of data. Only the biggest companies had the development capacity and budgets for big data. But the continuing decline in the costs of data storage, bandwidth, and computing means that big data is quickly becoming a useful and affordable tool for medium-sized and even some small companies. Big data analytics are already becoming the basis for many new, highly specialized business models.

2. Cloud Computing

The term cloud computing refers to computing in which tasks are performed by a "cloud" of servers, typically via the Internet (Fig. 4-5). This type of network has been used for several years to create the supercomputer-level power needed for research and other power-hungry applications, but it was more typically referred to as grid computing in this context. Today, cloud computing typically refers to accessing Web-based applications and data using a personal computer, mobile phone, or any other Internet-enabled device. The concept of cloud computing is that apps and data are available any time, from anywhere, and on any device. For example, you use cloud computing capabilities when you store or access documents, phones, videos, and other media online; use programs and apps online (i.e., e-mail, productivity, games, etc.); and share ideas, opinions, and content with others online (i.e., social networking sites).

Fig. 4-5 Cloud computing

The name cloud computing was inspired by the cloud symbol that is often used to represent the Internet in flow charts and diagrams. Clouds can be classified as public, private or hybrid. Businesses often use applications available in the public cloud; they also frequently create a private cloud just for data and applications belonging to their company.

Cloud computing relies on sharing of resources to achieve coherence and economies of scale, similar to a utility (like the electricity grid) over a network. At the foundation of cloud computing is the broader concept of converged infrastructure and shared services.

Cloud computing also focuses on maximizing the effectiveness of the shared resources. Cloud resources are usually not only shared by multiple users but are also dynamically reallocated per demand. This can work for allocating resources to users. For example, a cloud computer facility that servers European users during European business hours with a specific application (e.g. e-mail) may reallocate the same resources to serve North American users during North America's business hours with a different application (e.g., a Web server). This approach should maximize the use of computing power, thus reducing environmental damage as well since less power, air conditioning, rack space, etc. are required for a variety of functions. With cloud computing, multiple users can access a single server to retrieve and update their data without purchasing licenses for different applications.

Cloud computing providers offer their services according to three fundamental models: Infrastructure as a Service (IaaS), Platform as a Service (PaaS) and Software as a Service (SaaS).

Infrastructure as a Service like Amazon Web Services provides virtual server instances with unique IP addresses and blocks of storage on demand. Customers use the provider's application program interface (API) to start, stop, access and configure their virtual servers and storage. In the enterprise, cloud computing allows a company to pay for only as much capacity as is needed, and bring more online as soon as required. Because this pay-for-what-you-use model resembles the way electricity, fuel and water are consumed; it is sometimes referred to as utility computing.[4]

Platform as a Service in the cloud is defined as a set of software and product development tools hosted on the provider's infrastructure. Developers create applications on the provider's platform over the Internet. PaaS providers may use APIs, website portals or gateway software installed on the customer's computer. GoogleApps are examples of PaaS. Some providers will not allow software created by their customers to be moved off the provider's platform.

In the Software as a Service cloud model, the vendor supplies the hardware infrastructure, the software product and interacts with the user through a front-end portal. SaaS is a very broad market. Services can be anything from Web-based email to inventory control and database processing. Because the service provider hosts both the application and the data, the end user is free to use the service from anywhere.[5]

Key Words

affordable	付得起的
anomalies	异常，反常，不规则
billion	十亿，大量
decline	下降，衰退
drug	药物，毒品
epidemics	流行病，泛滥
exabyte	10^{18} 字节
license	许可证，执照，特许
petabyte	10^{15} 字节
phenomena	现象

proliferate	激增,扩散
rack	支架,行李架
reallocated	再分配,再指派
sophisticated	复杂的,精致的
trillion	万亿,兆
Twitter	推特,一个网站名称

Notes

[1] According to the IDC technology research firm, data is more than doubling every two years, so the amount of data available to organizations is skyrocketing.

译文:根据IDC技术研究公司的调研,数据每两年都会增长一倍以上,因此对组织而言可用的数据量不断激增。

本句中,"According to the"是条件状语,"so the amount of…"作结果状语。

[2] No single site server configuration offers sufficient capacity for the enormous amounts of data derived from sites such as Microsoft.com or FedEx.com, which serve global communities of customers, with transactions numbering in the millions.

译文:诸如Microsoft.com或FedEx.com这些网站,它们为全球的客户通信服务,有数以百万计的交易编号,对于从它们得到的巨大量的数据,没有单个的网站服务器配置能够提供足够的能力。

本句中"derived from…"作定语,"which"引导的是非限定性定语从句。

[3] Large Internet commerce companies such as Amazon and Taobao use big data analytics to predict buyer activity as well as to understand warehousing requirements and geographic positions.

译文:一些大的互联网商务公司,例如亚马逊和淘宝,使用大数据分析来预测买家活动以及了解仓储需求和地理定位。

本句中的"Amazon and Taobao"作主语的同位语,"to predict buyer…"是目的状语。

[4] Because this pay-for-what-you-use model resembles the way electricity, fuel and water are consumed; it is sometimes referred to as utility computing.

译文:因为这种"支付你所用的"的模式很像消费电、燃料和水的方式;它有时称为公共设施计算。

本句中的"pay-for-what-you-use"可以译为"支付你所用的","Because this…"是原因状语从句。

[5] Because the service provider hosts both the application and the data, the end user is free to use the service from anywhere.

译文:因为服务提供者既拥有应用程序,又拥有数据,因此终端用户可以自由地从任何地方使用该服务。

本句由because引导原因状语从句。

4.3.2 Exercises

1. Translate the following phrases into English or Chinese

(1) Internet-enabled device

(2) nontraditional data

（3）inventory control
（4）virtual server instance
（5）enterprise control language
（6）云计算
（7）私有云
（8）公有云
（9）移动电话
（10）大数据

2. Identify the following to be True or False according to the text

（1）The largest companies making the most sophisticated industrial equipment are now able to go from merely detecting equipment failures to predicting them.

（2）The term cloud computing refers to computing in which tasks are performed by a "cloud" of servers, typically via the Internet.

（3）Big data usually refers to data in the TB and PB range.

（4）Businesses are interested in big data because they contain more patterns and interesting anomalies than smaller data sets

（5）GoogleApps are examples of SaaS.

（6）Many big data applications were created to help Web companies struggling with unexpected volumes of data.

3. Reading Comprehension

（1）Cloud resources are usually not only shared by _____ but are also dynamically reallocated per demand.
　　A. multiple users
　　B. single user
　　C. big data
　　D. cloud computing

（2）_____ in the cloud is defined as a set of software and product development tools hosted on the provider's infrastructure.
　　A. PaaS
　　B. IaaS
　　C. SaaS
　　D. BaaS

（3）The enormous growth of data has coincided with the development of the "cloud", which is really just a new alias for the _____.
　　A. global Intranet
　　B. global Extranet
　　C. global Internet
　　D. local network

(4) Big data analysis requires a cloud-based solution—a networked approach capable of keeping up with the accelerating volume and speed of _____.
   A. data transmission
   B. information
   C. data
   D. data storage

(5) In the _____ cloud model, the vendor supplies the hardware infrastructure, the software product and interacts with the user through a front-end portal.
   A. BaaS
   B. PaaS
   C. IaaS
   D. SaaS

4.3.3 Reading Material

Top Technologies of Computer Science

Nanotechnology

Nanotechnology involves the creation of devices and material at the nanometer level, and therefore nanoelectronics is a natural progression of microelectronics. Just as microelectronics led to the creation of microchip or integrated circuit, instead of microchip, nanoelectronics could conceivably lead to the "nanochip". On June 26, 2007, IBM announced the second generation of its top supercomputer. Blue Gene is the world's fastest computer, at 360 TFLOPS. But the chips Power 5.6 what it used are still not a "nanochip".

Component-based development

Component-based development (CBD) will revolutionize the software industry. Unlike recent trends such as objects and client/server, it is not just another favor of distributed computing, but an extensible architecture to support all life cycle computing metaphor, including design, development and deployment. Because of its high levels of reuse and interoperability, CBD will influence every dimension of application composition, including all types of clients, application servers and database server, and will have a profound impact of application development.

CBD's predecessor was object-oriented development. CBD redefines objects in the context of a standardized infrastructure for interoperability frameworks for the construction and frameworks and assembly of application, and pre-built components that subscribe to component infrastructure and frameworks. This infrastructure, along with the availability CBD frameworks that enable component design, construction and assembly (rather than just bring an environment for visual programming), will forever change how application are developed. CBD can be divided into three stages: monolithic, distributed and persistent overtime, certain standards will become pervasive, core service will become commercial, and the infrastructural dimension of CBD will ensure that portability and interoperability are met.

Intelligent robots

Most existing robots are far from being intelligent, and they cannot hear, see, and feel. Existing robots can't adapt themselves to changes in its work plan and in work environment. It can't autonomously alter its action plan to fit with the changing situation, but wait for human intervention or run into chaos. Imagine the danger and audacious guy will face a complex, rapidly changing environment. These tell why we need intelligent robots.

An intelligent robot is basically one that must be capable of sensing its surroundings and possess intelligence enough to respond to a changing environment in much the same way as we do. Such abilities require the direct application of sensory perception and artificial intelligence.

Much research in robotics has been and is still concerned with how to equip robot with visual sensors "eyes" and tactile sensors "fingers". Artificial intelligence will enable the robot to respond to and adapt to changes in its tasks and in its environment, and to reason and make decision in reaction to those changes. In this sense, an intelligent robot can be described as one that is equipped with various sensors, possessed sophisticated signal processing capabilities and intelligence for making decision, in addition to general mobility.

Ubiquitous computing

Ubiquitous computing gives us tools to manage information easily. Information is the new currency of the global economy. We increasingly rely on the electronic creation, storage, and transmittal of personal, financial, and other confidential information, and demand the highest security for all these transactions. We require complete access to time-sensitive data, regardless of physical location. We expect devices—personal digital assistants, mobile phones, office PCs and home entertainment systems—to access that information and work together in one seamless, integrated system.

Ubiquitous computing aims to enable people to accomplish an increasing number of personal and professional transactions using a new class of intelligent and portable devices. It gives people convenient access to relevant information stored on powerful networks, allowing them to easily take action anywhere, anytime.

4.4 专业英语应用模块

4.4.1 The Interview Questions

面试是公司挑选职工的一种重要方法，也是你面对未来老板的一个重要过程。面试给公司和应聘者提供了进行双向交流的机会，能使公司和应聘者之间相互了解，从而双方都能够更准确地做出聘用与否、受聘与否的决定。

被面试官一眼相中的秘诀在于在面试中回答问题时，找到自信与谦虚之间的平衡。面试就好比是一场考试，既测试每个人的能力，也测试每个人的心理素质和临场发挥。因此，要成功面试，首先要充满信心，其次要准备好与应聘岗位相关的专业知识与业务技能，准备当天可能用到的个人资料或作品。招聘者可能会先评价一个求职者的衣着、外表、仪态及行为举止，也可能会对求职者的专业知识、口才、谈话技巧等做综合性的考核。面试时切忌伪装和掩饰，一定要展现自己的真实实力和真正的性格。

下面是英语面试中的10个常见问题，面试者需要提前做好书面准备。

1. Tell me about yourself.
简要介绍你自己。
2. Why are you interested in this position?
你为什么对这份工作感兴趣?
3. Why do you want to work for us?
你为什么想来我们公司工作?
4. What do you know about our company?
你想了解哪些关于我们公司的信息?
5. What are your strengths?
你的优势是什么?
6. What is your biggest weakness?
你最大的弱点是什么?
7. What are your interests?
你有哪些兴趣爱好?
8. What are your short and long term goals?
你的短期和长期目标是什么?
9. Why are you leaving your present job?
你为什么离开现在的公司?
10. How would you evaluate your present firm?
如何评价你现在的公司?
11. Why do you feel you are right for this position?
为什么你认为自己适合这个职位?
12. Why should we hire you?
我们为什么聘用你?
13. How long would you stay with us?
你能在本公司工作多长时间?
14. Why should I hire you over the other candidates I am interviewing?
我为什么要从这么多的应聘者中选择你呢?
15. What are your compensation expectations?
你对于报酬有什么样的期望?
16. How much are you looking for?
你期待的薪金是多少?
17. How do you prioritize when you are given too many tasks to accomplish?
你怎样在一堆根本做不完的工作任务中区分轻重缓急?

4.4.2 Resume

简历是有针对性的自我介绍的一种规范化、逻辑化的书面表达。简历是对个人学历、经历、成绩、特长及其他有关情况所作的简明扼要的书面介绍。成功的简历就是一件营销武器,它向未来的雇主表明自己拥有能够满足特定工作要求的技能、态度、资质和自信。

简历内容不要太挤太满,但也不要太空太稀,一般在一页或两页纸之内。简历上应提供客观的证明或者佐证资历、能力的事实和数据。一份适合职位要求、翔实和打印整齐的简历可以有效地获得与聘用单位面试的机会。

简历中的基本项目应包括如下几项。

1．个人基本信息

应根据求职的情况来列出自己的姓名、性别、年龄、国籍、政治面貌、学校、系别、专业、婚姻状况、健康状况、身高、爱好与兴趣、家庭住址、电话号码等。

示例如下：

> Wang Xiaolin
> Computer Science Department
> Shanghai Industrial University
> Shanghai, 200010, P.R.China
>
> Date of Birth: July 13, 1993　　Place of Birth: Shanghai, China
> Sex: Male　　　　　　　　　　　Nationality: Chinese
> Marital status: Single　　　　　　Health: Excellent
> Telephone: 021-64152878　　　　E-mail: Wangxiaolin@sohu.com

2．学历

学历主要指高中或高中以上的学历。应写明学校、专业名称，起止期间，并列出所学主要课程及学习成绩，在学校和班级所担任的职务，在校期间所获得的各种奖励和荣誉等。学历的排列顺序由高到低。

示例如下：

> Education
>
> 2011—2015　Computer Science
> 　　　　　　Shanghai Industrial University, Shanghai, China
> 2008—2011　No.2 middle school of Shanghai, Shanghai, China

3．经历

经历为工作资历情况，在简历中非常重要。若有工作经验，最好详细列明，首先列出最近的资料，后详述曾工作单位、日期、职位、工作性质等。经历的时间顺序一般为由近及远。

示例如下：

> Experience
>
> 2014—present　Sunlight Software company, Shanghai, China. Originally worked at a programmer, promoted to the position of project manager in 2015.
> 　　　　　　　Responsibilities including meeting customers, planning project, arranging job and checking result.
> 2011—2015　　Major in Computer Science. Being the computer science department chairman for two years.

PART 5 第 5 章

Computer Network Technology

5.1 Computer Network Basics

5.1.1 Text

Networking arose from the need to share data in a timely fashion. Personal computers are wonderful business tools for producing data, spreadsheets, graphics, and other types of information, but do not allow you to quickly share the data you have produced. Without a network, the documents have to be printed out so that others can edit them or use them.[1] At best, you can give files on floppy disks to others to be copied to their computers. If others make changes to the document there is no way to merge the changes. This was, and still is, called working in a stand-alone environment.

If the user were to connect his computer to other computers, he could share data and devices on other computers, including high-quality printers.[2] A group of computers and other devices connected together is called a network, and the technical concept of connected computers sharing resources is called networking (Fig. 5-1).

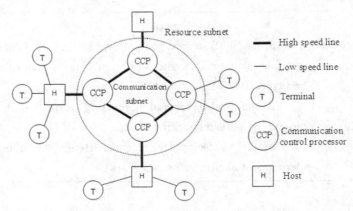

Fig. 5-1 Composition of computer network

Computers that are part of a network can share data, messages, graphics, printers, fax machines, modems, CD-ROMs, hard disks, and other data storage equipment.

1. Network Structure

Networks come in a variety of sizes and complexities. Generally, the bigger, the more complex, and the more difficult to manage.

(1) Local Area Network

A Local Area Network(LAN) is a homogenous entity. Every packet placed on the network is seen by every "node" (server or workstation), but each node processes only the packets addressed to it, except for network test equipment and hacker tools, which read any packets they please. All nodes on a LAN may be tapped into a single cable (old coax Ethernet) or wired to a central "hub".

(2) Bridged LAN

A bridged LAN network is a LAN that got too big and a "bridge" had to be inserted to cut down the traffic. A bridge knows what network addresses are on each side of it, and passes only traffic addressed to the other side, thus "segmenting" the network.

(3) Switched LAN

A switched LAN has a very fast, multi-port bridge called a "switch" or "switching hub". A switched LAN may be so fine grained it has only one node on each port, reducing traffic on each cable as far as it can be reduced.

(4) Campus Area Network

A Campus Area Network (CAN) will have at least one "backbone" to which two or more LANs are connected through bridges, switches or routers.

(5) Routed Network

A routed network is one where individual LANs are connected through routers, devices much smarter than bridges or switches. A router knows where stuff is, even if it's several routers away.[3] It keeps this information in a "routing table". It may know several "routes" to the same destination and may be smart enough to figure out which route is fastest.

(6) Wide Area Network

A Wide Area Network (WAN) is a network that passes over links not owned by the owner of the network. Generally these links are owned by a telephone company or some other common carrier. The Internet is a huge worldwide WAN consisting of thousands of LAN and millions of routers and servers (Fig. 5-2).

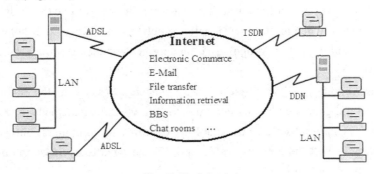

Fig. 5-2 Internet

2. Network Administrator

Network administrator has full rights and permissions to all resources on a network. The administrator is usually responsible for installing, managing, and controlling servers and networking components. Administrators can also modify the properties of user accounts and the membership of groups, create and manage printers, install printers, share resources and assign permissions to those resources. Database administrator is responsible for programming and maintaining a large multi-relational database in a networked environment and facilitating direct access to the database by individuals on the network. Workgroup manager is responsible for solving problems, implementing standards and solutions, reviewing performance, and facilitating the efficiency of a specific group of individuals who are connected to larger network environment. Support staff is responsible for providing technical assistance to the system administrator in large complex network environments, and providing routine problem solving and spot training to end-users. Maintenance contractor is responsible for hardware repairs and upgrades. Webmaster is responsible for implementing and maintaining the content and style of the company's Internet site, keeping the information accurate, up-to-date, and interesting.

3. Topologies

The physical topology describes the layout of the network, just like a map shows the layout of various roads, and the logical topology describes how the data is sent across the network or how the cars are able to travel (the direction and speed) at every road on the map. Logical topology is the method used to pass the information between the computers.

(1) Star network

A computer network with a star network topology in its simplest form, consists of one central, or hub computer which acts as a router to transmit messages between connected computers by a store-and-forward or switching system. A hierarchical extension of the star topology allows each node connected by a hub to in turn play the role of a hub for a disjoint set of leaf nodes. In this case multiple routes may exist between any two given nodes of the network.

(2) Grid network

A grid network is a kind of computer network consisting of a number of computer systems connected in a grid topology.[4] In a regular grid topology, each node in the network is connected with two neighbors along one or more dimensions. If the network is one-dimensional, and the chain of nodes is connected to form a circular loop, the resulting topology is known as a ring. Network systems such as FDDI use two counter-rotating token-passing rings to achieve high reliability and performance. In general, when an n-dimensional grid network is connected circularly in more than one dimension, the resulting network topology is an annulus, and the network is called "annular".

(3) Bus network

With the bus topology, all workstations are connected directly to the main backbone that carries the data. Traffic generated by any computer will travel across the backbone and be received by all workstations. This works well in a small network of 2~5 computers, but as the number of computers increases so will the network traffic and this can greatly decrease the performance and available bandwidth of your network.

In bus scheme workstations are stung along a cable (usually coax) like beads. Break it anywhere and the whole network goes down. Very unreliable and difficult to diagnose is because there are so many connectors and possible points of failure.

（4）Ring network

In ring network, the nodes are connected in a closed loop, or ring. Messages in a ring network pass in one direction, from node to node. As a message travels around the ring, each node examines the destination address attached to the message. If the address is the same as the address assigned to the node, the node accepts the message; otherwise, it regenerates the signal and passes the message along to the next node in the circle. Actually, there are usually two cables so adapters can loop back to remark the ring if there is a break.[5] In reality, most rings are virtual, and the actual wiring looks just like a star pattern.

（5）Hybrid network

In the hybrid topology, two or more topologies are combined to form a complete network. For example, a hybrid topology could be the combination of a star and bus topology. These are also the most common in use.

In a star-bus topology, several star topology networks are linked to a bus connection. In this topology, if a computer fails, it will not affect the rest of the network. However, if the central component, or hub, that attaches all computers in a star, fails, then you have big problems since no computer will be able to communicate.

In the star-ring topology, the computers are connected to a central component as in a star network. These components, however, are wired to form a ring network. Like the star-bus topology, if a single computer fails, it will not affect the rest of the network. By using token passing, each computer in a star-ring topology has an equal chance of communicating. This allows for greater network traffic between segments than in a star-bus topology.

Key Words

accurate	正确的，精确的
annulus	环
backbone	骨干，支柱
contractor	合同人，承包商
diagnose	诊断
fashion	风尚，时尚
hierarchical	分层的
hybrid	混合的
merge	使合并
neighbor	邻居，邻接
permission	许可，允许
regular	常规的，合格的
router	路由器
spreadsheet	电子表格
token	令牌，记号

topology	拓扑，结构
upgrade	升级，上升
unreliable	不可靠的

Notes

[1] Without a network, the documents have to be printed out so that others can edit them or use them.

译文：没有网络，就必须将文档打印出来才能供他人编辑或使用。

本句的"Without a network"是条件状语，"so that"引导的是目的状语从句。

[2] If the user were to connect his computer to other computers, he could share data and devices on other computers, including high-quality printers.

译文：如果用户能把他的计算机与其他计算机连在一起的话，他就可以共享其他计算机上的数据和设备，包括高性能的打印机。

本句由"If"引导条件状语从句，"including high-quality printers"是补语。

[3] A router knows where stuff is, even if it's several routers away.

译文：路由器知道信息的位置，即使要经过几个路由器的距离。

本句的"where stuff is"是宾语从句，"even if it's several routers away"是让步状语从句。

[4] A grid network is a kind of computer network consisting of a number of computer systems connected in a grid topology.

译文：网状网络是一种计算机网络，它以网状拓扑结构将许多计算机系统连接在一起。

本句的"consisting of a number of computer systems"是现在分词短语，作定语修饰"network"，而"connected in a grid topology"是过去分词短语作定语，修饰"computer systems"。

[5] Actually, there are usually two cables so adapters can loop back to remark the ring if there is a break.

译文：实际上，通常会有两条电缆，这样适配器就会在网络中出现故障时，可以沿环路返回重新构成环。

本句的"so adapters can loop back to remark the ring"是目的状语，而"if there is a break"是条件状语。

5.1.2　Exercises

1. Translate the following phrases into English or Chinese

（1）hybrid topology

（2）network traffic

（3）next node

（4）campus area network

（5）routed network

（6）网络管理员

（7）局域网

（8）网络环境

（9）物理拓扑

（10）逻辑拓扑

2. Identify the following to be True or False according to the text

(1) A bridge knows what network addresses are on each side of it.
(2) A LAN is a homogenous entity.
(3) A LAN is a network that passes over links not owned by the owner of the network.
(4) A routed LAN has a very fast, multi-port bridge called a "switch" or "switching hub".
(5) Network programmer has full rights and permissions to all resources on a network.
(6) In a star-bus topology, several star topology networks are linked to a bus connection.

3. Reading Comprehension

(1) Computers that are part of a _____ can share data, messages, graphics, printers, fax machines, modems, CD-ROMs, hard disks, and other data storage equipment.
   A. system
   B. unit
   C. structure
   D. network

(2) A switched _____ may be so fine grained it has only one node on each port, reducing traffic on each cable as far as it can be reduced.
   A. Router
   B. MAN
   C. WAN
   D. LAN

(3) _____ is responsible for programming and maintaining a large multi-relational database in a networked environment.
   A. Administrator
   B. Data processor
   C. Database administrator
   D. Programmer

(4) With the _____, two or more topologies are combined to form a complete network.
   A. bus topology
   B. ring topology
   C. hybrid topology
   D. network topology

(5) By using _____, each computer in a star-ring topology has an equal chance of communicating.
   A. signal
   B. token passing
   C. pointer
   D. counter

5.1.3 Reading Material
Optical Communication

A communication system transmits information from one place to another, whether separated by a few kilometers or by transoceanic distances. Information is often carried by an electromagnetic carrier wave whose frequency can vary from a few megahertz to several hundred terahertzs. Optical communication systems use high carrier frequencies in the visible or near-infrared region of the electromagnetic spectrum. They are sometimes called light-wave systems to distinguish them from microwave systems, whose carrier frequency is typically smaller by five orders of magnitude (1GHz). Fiber-optic communication systems are light-wave systems that employ optical fibers for information transmission. Such systems have been deployed worldwide since 1980 and have indeed revolutionized the technology behind telecommunications.

Microwave radio is used in broadcasting and telecommunication transmissions because, due to their short wavelength, highly directional antennas are smaller and therefore more practical than they would be at longer wavelengths (lower frequencies). There is also more bandwidth in the microwave spectrum than in the rest of the radio spectrum; the usable bandwidth below 300MHz is less than 300MHz while many GHz can be used above 300GHz. typically, microwaves are used in television news to transmit a signal from a remote location to a television station from a specially equipped van.

Optical communication systems differ in principle from microwave systems only in the frequency range of the carrier wave used to carry the information. The optical carrier frequencies are typically 200THz, in contrast with the microwave carrier frequencies are 1GHz. An increase in the information capacity of optical communication systems by a factor of up to 10 000 is expected simply because of such high carrier frequencies used for light-wave systems. This increase can be understood by noting that the bandwidth of the modulated carrier can be up to a few percent of the carrier frequency. Taking 1% as the limiting value, optical communication systems have the potential of carrying information at bit rates 1Tbps. It is this enormous potential bandwidth of optical communication systems that is the driving force behind the worldwide development and deployment of light-wave systems. Current state-of-the-art systems operate at bit rates 10Gbps, indicating that there is considerable room for improvement.

The application of optical fiber communications is in general possible in any area that requires transfer of information from one place to another. However, fiber-optic communication systems have been developed mostly for telecommunications applications. This is understandable in view of the existing worldwide telephone networks which are used to transmit not only voice signals but also computer data and fax messages. The telecommunication applications can be broadly classified into two categories, long-haul and short-haul, depending on whether the optical signal is transmitted over relatively long or short distances compared with typical intercity distances (100km). Long-haul telecommunication systems require high-capacity trunk lines and benefit most by the use of fiber-optic light-wave systems. Indeed, the technology behind optical fiber communication is often driven by long-haul applications.

Each successive generation of light-wave systems is capable of operating at higher bit rates and over longer distances. Periodic regeneration of the optical signal by using repeaters is still required for most long-haul systems. However, more than an order of magnitude increase in both the repeater spacing and the bit rate compared with those of coaxial systems has made the use of light-wave system very attractive for long-haul applications. Furthermore, transmission distances of thousands of kilometers can be realized by using optical amplifiers. A large number of transoceanic light-wave systems have already been installed to create an international fiber-optic network.

Short-haul telecommunication applications cover intra-city and local-loop traffic. Such systems typically operate at low bit rates over distances of less than 10km. The use of single-channel light-wave systems for such applications is not very cost-effective, and multichannel networks with multiple services should be considered. The concept of a broadband integrated-services digital network requires a high-capacity communication system capable of carrying multiple services. The asynchronous transfer mode (ATM) technology also demands high bandwidths. Only fiber-optic communication systems are likely to meet such wideband distribution requirements.

5.2 Information Security

5.2.1 Text

Security threats to a network can be divided into those that involve some sort of unauthorized access and all others. Once someone gains unauthorized access to the network, the range of things they can do is large. Some people are just interested in the challenge of breaking through the security and have no interest in doing anything further. They may, however, choose to monitor network traffic, called eavesdropping, for the purpose of learning something specific and perhaps disclosing it to others, or for the purpose of analyzing traffic patterns, traffic analysis could lead to the observation. The types of unauthorized access we usually think of, however, are the active security attacks whereby, after someone gains unauthorized access to the network, they take some overt action. Active attacks include altering message contents, masquerading as someone else, denial of service, and planting viruses.

Management's responsibility

The network security policy is management's statement of the importance and their commitment to network security. The policy needs to describe in general terms what will be done, but does not deal with the way the protection is to be achieved. Writing the policy is complex, because in reality, it is normally a part of a broader document: the organization's information security policy. The network security policy needs to clearly state management's position about the importance of network security and the items that are to be protected. Management must understand that there is no such thing as a perfectly secure network.

Furthermore, network security is a constantly moving target because of advances in technology and the creativity of people who would like to break into a network or its attached computers. Measures put in place to minimize security risks today will need to be upgraded in the future, and upgrades usually have a price attached, for which management will have to pay.

Simply writing the policy does not put the practices, procedures, or software in place to improve the security situation. That requires follow through and communication with all employees so that they understand the emphasis and importance senior management is placing on security.[1] Management, at all levels, needs to support the policy and periodically reinforce it with employees in various ways. IT and network staff may need to install additional hardware, software, and procedures to perform automated security checking.

Attack prevention

Different security mechanisms can be used to enforce the security properties defined in a given security policy. Depending on the anticipated attacks, different means have to be applied to satisfy the desired properties. Attack prevention is a class of security mechanisms that contains ways of preventing or defending against certain attacks before they can actually reach and affect the target. An important element in this category is access control, a mechanism which can be applied at different levels such as the operating system, the network, or tile application layer.

Access control limits and regulates the access to critical resources. This is done by identifying or authenticating the party that requests a resource and checking its permissions against the rights specified for the demanded object. It is assumed that an attacker is not legitimately permitted to use the target object and is therefore denied access to the resource. As access is prerequisite for an attack, any possible interference is prevented.

The most common form of access control used in multi-user computer systems are access control lists for resources that are based on the user and group identity of the process that attempts to use them.[2] The identity of a user is determined by an initial authentication process that usually requires a name and a password. The login process retrieves the stored copy of the password corresponding to the user name and compares it with the presented one. When both match, the system grants the user the appropriate user and group credentials. When a resource should be accessed, the system looks up the user and group in the access control list and grants or denies access as appropriate.[3] An example of this kind of access control can be found in the UNIX file system, which provides read, write and execute permissions based on the user and group membership. In this example, attacks against files that a user is not authorized to use are prevented by the access control part of the file system code in the operating system.

A firewall is an important access control system at the network layer. The idea of a firewall is based on the separation of a trusted inside network of computers under single administrative control from a potential hostile outside network. The firewall is a central choke point that allows enforcement of access control for services that may run at the inside or outside. The firewall prevents attacks from the outside against the machines in the inside network by denying connection attempts from unauthorized parties located outside, in addition, a firewall may also be utilized to prevent users behind the firewall from using certain services that are outside.[4]

Attack avoidance

Security mechanisms in this category assume that an intruder may access the desired resource but the information is modified in a way that makes it unusable for the attacker. The information is preprocessed at the sender before it is transmitted over the communication channel and post-processed at the receiver.[5] While the information is transported over the communication channel, it resists attacks by being nearly useless for an intruder. One notable exception is attacks against the availability of the information, as an attacker could still interrupt the message. During the processing step at the receiver, modifications or errors that might have previously occurred can be detected (usually because the information can not be correctly reconstructed). When no modification has taken place, the information at the receiver is identical to the one at the sender before the preprocessing step.

The most important member in this category is cryptography, which is defined as the science of keeping message secure. It allows the sender to transform information into what may seem like a random data stream to an attacker, but can be easily decoded by an authorized receiver.

The original message is called plaintext. The process of converting this message through the application of some transformation rules into a format that hides its substance is called encryption. The corresponding disguised message is denoted as ciphertext, and the operation of turning it back into plaintext is called decryption. It is important to notice that the conversion from plain to ciphertext has to be lossless in order to be able to recover the original message at the receiver under all circumstances.

The transformation rules are described by a cryptographic algorithm. The function of this algorithm is based on two main principles: substitution and transposition. In the case of substitution, each element of the plaintext (e.g., bit, block) is mapped into another element of the used alphabet. Transposition describes the process where elements of the plaintext are rearranged. Most systems involve multiple steps of transposition and substitution to be more resistant against cryptanalysis. Cryptanalysis is the science of breaking the cipher, i.e., discovering the substance of the message behind its disguise.

The most common attack, called known plaintext attack, is executed by obtaining ciphertext together with its corresponding plaintext. The encryption algorithm must be so complex that even if the code breaker is equipped with plenty of such pairs, it is infeasible for here to retrieve the key. An attack is infeasible when the cost of breaking the cipher exceeds the value of the information, or the time it takes to break it exceeds the lifespan of the information itself.

Given pairs of corresponding cipher and plaintext, it is obvious that a simple key guessing algorithm will succeed after some time. The approach of successively trying different key values until the correct one is found is called brute force attack because no information about the algorithm is utilized. In order to be useful, it is a necessary condition for an encryption algorithm that brute force attacks are infeasible.

Key Words

anticipate	预料，预测
brute	暴力的，粗暴的
choke	要塞，拥塞
ciphertext	密文
creativity	创造力，创新
cryptographic	加密的，密码的

decode	译码
denial	否认，拒绝
denote	表示，意味着
disguise	伪装，借口
eavesdrop	偷听，窃听
encryption	加密
hostile	敌人，敌对分子
interference	冲突，干扰
intruder	闯入者，侵入者
lossless	无损的
observation	观察，观测
overt	公开的，公然的，明显的
plaintext	明文
separation	分离，隔离，隔开
upgrade	升级，提高

Notes

[1] That requires follow through and communication with all employees so that they understand the emphasis and importance senior management is placing on security.

译文：这需要后续通过与所有员工的沟通，使他们认识到高层管理人员在安全上布置的重点和重要性。

本句的"so that"引导目的状语从句。

[2] The most common form of access control used in multi-user computer systems are access control lists for resources that are based on the user and group identity of the process that attempts to use them.

译文：在多用户计算机系统中，最常用的访问控制形式是资源访问控制清单。清单是建立在进程中想要使用这些资源的用户和群的身份基础上的。

本句的"used in multi-user computer systems"作定语，修饰"access control"，"that are based…"是修饰"access control lists"的定语。

[3] When a resource should be accessed, the system looks up the user and group in the access control list and grants or denies access as appropriate.

译文：当一个资源被请求访问时，系统会在访问控制清单中查找用户和群并适当授权访问或拒绝访问。

本句的"When"引导时间状语从句，"in the access control list"作定语。

[4] The firewall prevents attacks from the outside against the machines in the inside network by denying connection attempts from unauthorized parties located outside, in addition, a firewall may also be utilized to prevent users behind the firewall from using certain services that are outside.

译文：防火墙通过拒绝来自外部的未授权方的连接企图，从而防止来自外部的对内部网络中的机器的攻击。除此之外，防火墙还可以用来阻止墙内用户使用外部服务。

这是一个长句，由"in addition"连接起两个并列子句，"by denying connection attempts from unauthorized parties located outside"是方式状语，"behind the…"作定语。

[5] The information is preprocessed at the sender before it is transmitted over the communication channel and post-processed at the receiver.

译文： 信息在通信渠道中传播之前由发送者进行了预处理，并且接收者收到信息后也要进行处理。

本句采用被动语态，主语是 "The information"，谓语为 "preprocessed" 和 "post-processed"，构成并列句，"before it is transmitted over the communication channel" 是时间状语。

5.2.2 Exercises

1. Translate the following phrases into English or Chinese

（1）brute force attack
（2）unauthorized access
（3）cryptographic algorithm
（4）attack avoidance
（5）authentication process
（6）访问控制
（7）攻击预防
（8）网络安全策略
（9）随机数据流
（10）安全威胁

2. Identify the following to be True or False according to the text

（1）Measures put in place to minimize security risks today will need to be upgraded in the future.

（2）Different security mechanisms can be used to enforce the security properties defined in a given security policy.

（3）IT and network staff may not need to install additional hardware, software, and procedures to perform automated security checking.

（4）The identity of a user is determined by an initial authentication process that only requires a name.

（5）A password is an important access control system at the network layer.

（6）The most common attack, called known plaintext attack, is executed by obtaining ciphertext together with its corresponding plaintext.

3. Reading Comprehension

（1）When no modification has taken place, the information at the receiver is identical to the one at the sender before the _____ step.

  A. post-processing
  B. preprocessing
  C. processing
  D. first

（2）The approach of successively trying different key values until the correct one is found is called _____.

A. attack prevention
B. attack avoidance
C. network attack
D. brute force attack

(3) Cryptanalysis is the science of breaking the cipher, i.e., discovering the substance of the message behind its _____.
A. cover
B. disguise
C. truth
D. reality

(4) Active attacks include altering _____, masquerading as someone else, denial of service, and planting viruses.
A. message contents
B. data
C. statements
D. commands

(5) An attack is infeasible when the cost of breaking the cipher exceeds the value of the information, or the time it takes to break it exceeds the _____ of the information itself.
A. time
B. date
C. number
D. lifespan

5.2.3 Reading Material

Techniques for Internet Security

1. Network Firewalls

The purpose of a network firewall is to provide a shell around the network that will protect the systems connected to the network from various threats. The types of threats a firewall can protect against include:

- Unauthorized access to network resources—an intruder may break into a host on the network and again unauthorized access to files.
- Denial of service—an individual from outside of the network could, for example, send thousands of mail messages to a host on the net in an attempt to fill available disk space or load the network links.
- Masquerading—electronic mail appearing to have originated from one individual could have been forged by another with the intent to embarrass or cause harm.

A firewall can reduce risks to network systems by filtering out inherently insecure network services (Fig. 5-3). Network file system (NFS) services: for example, could be prevented from being used from outside of a network by blocking all NFS traffic to or from the network. This protects the individual hosts while still allowing the service, which is useful in a LAN environment, on the internal network. One way to avoid the problems associated with network computing would be to completely disconnect an organization's internal network from any other external system. This, of course, is not the preferred method. Instead what is needed is a way to filter access to the network while still allowing users access to the "outside world".

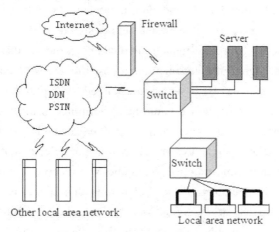

Fig. 5-3 Network structure example

In this configuration, the internal network is separated from external networks by a firewall gateway. A gateway is normally used to perform relay services between two networks. In the case of a firewall gateway, it also provides a filtering service that limits the types of information that can be passed to or from hosts located on the internal network. There are three basic techniques used for firewalls: packet filtering, circuit gateway, and application gateways. Often, more than one of these is used to provide the complete firewall service.

2. Digital Certificate and Authentication

Digital certificate is an identity card counterpart in the computer society. When a person wants to get a digital certificate, he generates his own key pair, gives the public key as well as some proof of his identification to the Certificate Authority (CA). CA will check the person's identification to assure the identity of the applicant. If the applicant is really the one "who claims to be", CA will issue a digital certificate, with the applicant's name, E-mail address and the applicant's public key, which is also signed digitally with the CA's private key. When A wants to send B a message, instead of getting B's public key, A now has to get B's digital certificate. A first checks the certificate authority's signature with the CA's public key to make sure it is a trustworthy certificate. Then A obtains B's public key from the certificate, and uses it to encrypt message and sends to B.

3. Digital Signature

For ages, special seals or handwritten signatures on documents have served as proof of authorship of, or agreement with, the contents of a document. Several attributes make the use of handwritten signatures compelling. These include the following.

- A signature is not forgeable and serves as proof that the signer deliberately signed the document.
- A signature is authentic and convinces the recipient that the signer deliberately signed the

document.
- A signature is not reusable. It is part of the document, and an unscrupulous person cannot transfer it to a different document.
- Once signed, a document is unalterable.
- A signature cannot be repudiated. Since the signature and the document are physical objects, the signer cannot later claim that he or she did not sign it.

A digital signature is an electronic signature that can be used to authenticate the identity of the sender of a message or the signer of a document, and possibly to ensure that the original content of the message or document that has been sent is unchanged. A digital signature is a string of bits attached to an electronic document, which could be a word processing file or an E-mail message. This bit string is generated by the signer, and it is based on both the document's data and the person's secret password. Digital signatures are easily transportable, cannot be imitated by someone else, and can be automatically time-stamped. Someone who receives the document can prove that the signer actually signed the document. If the document is altered, the signer can also prove that he did not sign the altered document. A digital signature can be used with any kind of message, whether it is encrypted or not, simply so that the receiver can be sure of the sender's identity and that the message arrived intact.

5.3 Wireless Networks

5.3.1 Text

Wireless technologies, in the simplest sense, enable one or more devices to communicate without physical connections—without requiring network or peripheral cabling. Wireless technologies use radio frequency transmissions as the means for transmitting data, whereas wired technologies use cables. Wireless technologies range from complex systems, such as wireless local area networks (WLAN) and cell phones to simple devices such as wireless headphones, microphones, and other devices that do not process or store information.[1] They also include infra-red (IR) devices such as remote controls, some cordless computer keyboards and mice, and wireless hi-fi stereo headsets, all of which require a direct line of sight between the transmitter and the receiver to close the link.

Wireless networks serve as the transport mechanism between devices and among devices and the traditional wired networks (enterprise networks and the Internet). Wireless networks are many and diverse but are frequently categorized into three groups based on their coverage range: wireless wide area networks (WWAN), WLANs, and wireless personal area networks (WPAN). WWAN includes wide coverage area technologies such as 2G cellular, cellular digital packet data (CDPD), global system for mobile communication (GSM), and Mobitex. WLAN includes 802.11, hyper LAN, and several others. WPAN, represents wireless personal area network technologies such as Bluetooth and IR. All of these technologies are "tetherless"—they receive and transmit information using electromagnetic waves. Wireless technologies use wavelengths ranging from the radio frequency (RF) band up to and above the IR band. The frequencies in the RF band cover a significant portion of the electromagnetic radiation spectrum, extending from 9 kilohertz, the lowest allocated wireless communications frequency, to thousands of gigahertz. As the frequency is increased beyond the RF spectrum, electromagnetic energy moves into the IR and then the visible spectrum.

WLANs allow greater flexibility and portability than do traditional wired local area networks. Unlike a traditional LAN, which requires a wire to connect a user's computer to the network, a WLAN connect computers and other components to the network using an access point device.[2] An access point communicates with devices equipped with wireless network adaptors; it connects to a wired Ethernet LAN via an RJ-45 port. Access point devices typically have coverage areas of up to 300 feet (approximately 100 meters). This coverage area is called a cell or range. Users move freely within the cell with their laptop or other network device. Access point cells can be linked together to allow users to even "roam" within a building or between buildings.

Ad hoc networks such as Bluetooth are networks designed to dynamically connect remote devices such as cell phones, laptops, and PDAs. These networks are termed "ad hoc" because of their shifting network topologies. Whereas WLANs use a fixed network infrastructure, ad hoc networks maintain random network configurations, relying on a master-slave system connected by wireless links to enable devices to communicate. In a Bluetooth network, the master of the piconet controls the changing network topologies of these networks. It also controls the flow of data between devices that are capable of supporting direct links to each other. As devices move about in an unpredictable fashion, these networks must be reconfigured on the fly to handle the dynamic topology. The routing that protocol Bluetooth employs allow the master to establish and maintain these shifting networks.

In recent years, wireless networking has become more available, affordable, and easy to use. Home users are adopting wireless technology in great numbers. On-the-go laptop users often find free wireless connections in places like coffee shops and airports.

Wireless network have many uses. For example, you can send E-mail, receive telephone calls and faxes and read remote documents in your travel. In addition, wireless networks are of great value to fleets of trucks, taxis, buses and repairpersons for keeping in contact with home. Although wireless LAN is easy to install, it also has some disadvantages. Typically they have a lower capacity, which is much slower than wired LAN. The error rates are often much higher too, and the transmissions from different computers can interface with one another.

If you are using wireless technology, or considering making the move to wireless, you should know about the security threats you may encounter. If you fail to secure your wireless network, anyone with a wireless-enabled computer within range of your wireless access point can hop a free ride on the Internet over your wireless connection.[3] The typical indoor broadcast range of an access point is 150-300 feet. Outdoors, this range may extend as far as 1 000 feet. So, if your neighborhood is closely settled, or if you live in an apartment or condominium, failure to secure your wireless network could potentially open your Internet connection to a surprising number of users. Doing so invites a number of problems:

- Service violations: You may exceed the number of connections permitted by your Internet service provider.

- Bandwidth shortages: Users piggybacking on your Internet connection might use up your bandwidth and slow your connection.
- Abuse by malicious users: Users piggybacking on your Internet connection might engage in illegal activity that will be traced to you.
- Monitoring of your activity: Malicious users may be able to monitor your Internet activity and steal passwords and other sensitive information.
- Direct attack on your computer: Malicious users may be able to access files on your computer, install spy ware and other malicious programs, or take control of your computer.

Wardriving is a specific kind of piggybacking. The broadcast range of a wireless access point can make Internet connections possible outside your home, even as far away as your street. Savvy computer users know this, and some have made a hobby out of driving through cities and neighborhoods with a wireless-equipped computer—sometimes with a powerful antenna—searching for unsecured wireless networks. This practice is nicknamed "wardriving". Wardrivers often note the location of unsecured wireless networks and publish this information on web sites. Malicious individuals wardrive to find a connection they can use to perpetrate illegal online activity using your connection to mask their identities. They may also directly attack your computer.

Wireless access points can announce their presence to wireless-enabled computers. This is referred to as "identifier broadcasting". In certain situations, identifier broadcasting is desirable. For instance, an Internet cafe would want its customers to easily find its access point, so it would leave identifier broadcasting enabled. When you use a wireless router or access point to create a home network, you trade wired connectivity for connectivity delivered via a radio signal.[4] Unless you secure this signal, strangers can piggyback on your Internet connection or, worse, monitor your online activity access files on your hard drive.

Many public access points are not secured, and the traffic they carry is not encrypted. This can put your sensitive communications or transactions at risk. Because your connection is being transmitted "in the clear", malicious users can use "sniffing" tools to obtain sensitive information such as passwords, bank account numbers, and credit card numbers.[5]

As is the case with unsecured home wireless networks, an unsecured public wireless network combined with unsecured file sharing can spell disaster. Under these conditions, a malicious user could access may directories and files you have allowed for sharing. Accessing the Internet via a public wireless access point involves serious security threats you should guard against. These threats are compounded by your inability to control the security setup of the wireless network. What's more, you are often in range of numerous wireless-enabled computers operated by people you don't know.

Key Words

abuse	欺骗
affordable	负担得起
Bluetooth	蓝牙
cellular	蜂窝的，多孔的
cordless	无绳的，不用电线的

coverage	范围，规模
disaster	灾害，天灾
diverse	多种多样的，不同的
fashion	方式，风格
headphone	耳机
inability	无能，无能为力
infrastructure	基站，基础设施
laptop	微型便携式（计算机）
fleet	船队，车队
malicious	有恶意的，蓄意的
nickname	诨名，绰号
repairperson	维修人员
roam	漫游
savvy	精明的，有见识的
unpredictable	无法预言的
violation	侵害，违犯
wavelength	波长
wireless	无线的，无线电的

Notes

[1] Wireless technologies range from complex systems, such as wireless local area networks (WLAN) and cell phones to simple devices such as wireless headphones, microphones, and other devices that do not process or store information.

译文：无线电技术的范围从复杂的系统如无线局域网和手机到简单的设备如无线耳机、麦克风以及其他不用来处理或储存信息的设备。

本句有两处由"such as"引导的同位语结构，"that do…"作定语。

[2] Unlike a traditional LAN, which requires a wire to connect a user's computer to the network, a WLAN connect computers and other components to the network using an access point device.

译文：与需要一根线把用户的计算机与网络连接的传统局域网不同，无线局域网使用一个接入点装置把计算机和其他组成部分连接到网络。

本句中"Unlike a traditional LAN"作状语，"which"引导的是非限定性定语从句，修饰"a traditional LAN"，"using an access point device"作方式状语。

[3] If you fail to secure your wireless network, anyone with a wireless-enabled computer within range of your wireless access point can hop a free ride on the Internet over your wireless connection.

译文：如果你不能确保你的无线网络的安全，任何拥有一台能激活无线网络的计算机的人只要在你的无线接口范围之内就能通过你的无线连接免费上互联网。

本句中"If"引导的是条件状语从句，"with a wireless-enabled computer within range of your wireless access point"作定语，修饰主语"anyone"。

[4] When you use a wireless router or access point to create a home network, you trade wired connectivity for connectivity delivered via a radio signal.

译文：当你使用一个无线路由器或接入点去建立一个家庭网络时，你用有线连接交换到由无线电信号传送的连接。

本句由"When"引导时间状语从句。

[5] Because your connection is being transmitted "in the clear", malicious users can use "sniffing" tools to obtain sensitive information such as passwords, bank account numbers, and credit card numbers.

译文：由于你的连接被畅通无阻地传送出去，恶意用户能够使用"嗅探"工具来获得敏感信息，诸如密码、银行账号以及信用卡号。

本句由"Because"引导原因状语从句，"such as"后面的部分作为"sensitive information"的同位语。

5.3.2 Exercises

1. Translate the following phrases into English or Chinese

（1）access point cell

（2）radio frequency transmission

（3）infra-red device

（4）service violation

（5）sensitive information

（6）电磁波

（7）网络技术

（8）无线技术

（9）恶意程序

（10）无线广域网

2. Identify the following to be True or False according to the text

（1）Wardrivers often note the location of secured wireless networks and publish this information on web sites.

（2）Wireless access points can announce their presence to wireless-enabled computers.

（3）In a Bluetooth network, the master of the piconet controls the changing network topologies of these networks.

（4）Many public access points are secured, but the traffic they carry is not encrypted.

（5）Wireless networks serve as the transport mechanism between devices and among devices and the traditional wired networks.

（6）Access point devices typically have coverage areas of up to 500 feet.

3. Reading Comprehension

（1）Unless you secure this signal, _____ can piggyback on your Internet connection or, worse, monitor your online activity access files on your hard drive.

A. administrators

B. strangers

C. receivers

D. senders

(2) The frequencies in the RF band cover a significant portion of the electromagnetic radiation spectrum, extending from _____, the lowest allocated wireless communications frequency, to thousands of gigahertz.
   A. 6 kilohertz
   B. 8 kilohertz
   C. 9 kilohertz
   D. 12 kilohertz

(3) If you are using wireless technology, or considering making the move to wireless, you should know about the _____ you may encounter.
   A. transmission problems
   B. Internet
   C. networks
   D. security threats

(4) Users piggybacking on your _____ might use up your bandwidth and slow your connection.
   A. Internet connection
   B. password
   C. username
   D. network

(5) In addition, _____ are of great value to fleets of trucks, taxis, buses and repairpersons for keeping in contact with home.
   A. wireless networks
   B. networks
   C. Internet
   D. Intranet

5.3.3 Reading Material

Internet Applications

● E-mail

E-mail refers to the transmission of messages over communications networks. E-mail uses the office memo paradigm in which a message contains a header that specifies the sender, recipients, and subject, followed by a body that contains the text of the message. To participate in E-mail, a person must assign a mailbox, which is in fact a storage area where messages can be placed.

An E-mail address is a string divided into two parts by the @ character (pronounced as "at"). The first part is a mailbox identifier, and the second stands for the name of the computer on which the mailbox resides. Mailbox identifiers are assigned locally, and only have significance on one computer. On some computer systems, the mailbox identifier is the same as a user's login account identifier; on other systems, the two are independent. The computer name in an E-mail address is a domain name.

● Search Tools and Methods

A search tool is a computer program that performs searches. A search method is the way a search tool requests and retrieves information from its Web site. A search begins at a selected search tool's Web site, reached by means of its address or URL. There are essentially four types of search tools, each of which has its own search method.

(1) A directory search tool searches for information by subject matter.

(2) A search engine tool searches for information through use of keywords and responds with a list of references.

(3) A directory with search engine uses both the subject and keyword search methods interactively as described above.

(4) A multi-engine search tool utilizes a number of search engines in parallel.

Each search engine has its own way of assigning relevance. There are three methods used in the indexing of a Web site database.

(1) Full text index

(2) Keyword index

(3) Person index

- A mobile Web

The mobile Web refers to access to the World Wide Web, i.e. the use of browser-based Internet services, from a handheld mobile device, such as a smart phone or a feature phone, connected to a mobile network or other wireless network.

Traditionally, access to the Web has been via fixed-line services on large-screen laptops and desktop computers. However, the Web is becoming more accessible by portable and wireless devices. An early 2010 ITU (International Telecommunication Union) report said that with the current growth rates, Web access by people on the go—via laptops and smart mobile devices—is likely to exceed Web access from desktop computers within the next five years. The shift to mobile Web access has been accelerating with the rise since 2007 of large multi-touch smart phones, and of multi-touch tablet computers since 2010. Both platforms provide better Internet access, screens, and mobile browsers—or application-based user Web experiences than previous generations of mobile devices have done. Web designers may work separately on such pages, or pages may be automatically converted as in Mobile Wikipedia.

The distinction between mobile Web applications (mobile apps) and native applications is anticipated to become increasingly blurred, as mobile browsers gain direct access to the hardware of mobile devices (including accelerometers and GPS chips), and the speed and abilities of browser-based applications improve. Persistent storage and access to sophisticated user interface graphics functions may further reduce the need for the development of platform-specific native applications.

Mobile Web access today still suffers from interoperability and usability problems. Interoperability issues stem from the platform fragmentation of mobile devices, mobile operating systems, and browsers. Usability problems are centered around the small physical size of the mobile phone form factors (limits on display resolution and user input/operating). Despite these shortcomings, many mobile developers choose to create apps using mobile Web.

Mobile Internet refers to access to the Internet via a cellular telephone service provider. It is wireless access that can hand off to another radio tower while it is moving across the service area. It can refer an immobile device that stays connected to one tower, but this is not the meaning of "mobile" here. Wi-Fi and other better methods are commonly available for users not on the move. Cellular base stations are more expensive to provide than a wireless base station that connects directly to an Internet service provider, rather than through the telephone system.

A mobile phone, such as a smart phone, that connects to data or voice services without going through the cellular base station is not on mobile Internet. A laptop with a broadband modem and a cellular service provider subscription that is traveling on a bus through the city is on mobile Internet.

A mobile broadband modem "tethers" the smart phone to one or more computers or other end user devices to provide access to the Internet via the protocols that those cellular telephone service providers may offer.

5.4　专业英语应用模块

5.4.1　Self-Introduction

自我介绍是向别人展示自己的一个重要手段，这个环节直接关系到你给别人的第一印象。自我介绍也是一种说服的手段和艺术。要想让面试官认可你，你必须明确地告诉他们你具有的能力与素质。因此，组织自我介绍内容时，既要表现出你优秀的特质，也要表达出你适合这个职位的相关信息。

自我介绍的时间不长，一般谈谈学历、个人基本情况、工作经历、社会实践、对于应聘职位和行业的看法等。在进行口语自我介绍前，面试者需要提前写好书面的自我介绍。

自我介绍的原则：
1. 开门见山，简明扼要，一般最好不要超过三分钟。
2. 实事求是，不可吹得天花乱坠。
3. 突出长处，但要与申请的职位有关。
4. 善于用具体生动的实例来证明自己，说明问题。
5. 介绍完后，要问面试官还想知道关于自己的什么事情。

自我介绍示例如下：

Dear Sir or Madam,

Thank you very much for reading my application and I am much honored to introduce myself here.

My name is Zhang Xiaoli, graduated from Shanghai Industrial University in 2015. I am 23 years old and I come from Shanghai. I am seeking an opportunity to work with ××× as an Engineer. My professional experience and my awareness of your unparalleled reputation have led me to want to work for your company.

I have a bachelor degree with a major in Electronic field. During the four year undergraduate study, my academic records kept distinguished. I was granted scholarship every semester.

I participated in many school activities, which widened my horizons and gave me many opportunities to do practical work in companies. All of that were very useful to my major study. During this period, I learned the values of teamwork and commitment, how to win, how to work hard, how to concentrate and focus on goals, and how to balance my time and priorities. The passing years offered me a good chance to give full play to my creativity, intelligence and diligence.

With a healthy body, with the solid professional knowledge, with the youthful passion, with the yearning for the future and the admiration of your company, I am eager to enter your company and make my share of contribution to it.

Thank you for your time.
Sincerely yours,
Zhang Xiaoli

5.4.2 Self-Recommendation

自荐信与你的履历表一样重要，一份好的自荐信能为你赢得一个面试机会。履历表告诉别人你的经历和你的技能，而自荐信告诉别人你能为雇佣者做些什么。

1．自我介绍和你是如何得知该职位的招聘信息

在介绍自我部分，可以用一句话简单介绍一下你自己。只要把最重要、也是与未来雇主最有关的信息写清楚就可以了。

示例如下：

I am the senior in college, major in computer science. I will graduate in July.

在信息来源部分，说明你对该公司有兴趣并想应聘他们空缺的职位。可以通过暗示你与公司雇员的亲属关系来表达你对公司的兴趣。如果你由一位朋友或者同事介绍给公司的，就在信中提起他们，因为招聘经理会感到有责任回复你的信。

示例如下：

I am glad to know your company is hiring website designers, a customer of your company recommended me.

2．自我推荐

简短地叙述自己的才能，特别是这些才能将满足公司的需要。不要在信中表示你会因聘用而受益多少，面对桌上一大堆履历表和许多空缺职位，招聘经理关心的不会是你的个人成就。陈述你所特有的将能为公司做出贡献的教育、技能和个性特征。该自荐信应促使招聘者想进一步阅读你的简历。

示例如下：

I was rewarded as my activity and creation, during I worked for Blue Sky Network company as a computer network maintainer.

3．致谢及进一步行动的要求

信的结尾部分不仅仅只是对你的雇主花时间读你的信表示感谢，还要表明你的下一步计划，不要让招聘者来决定，要自己采取行动，告诉招聘者怎样才能与你联系，打电话或者发邮件，但不要坐等电话，要表明如果几天内等不到他们的电话，你会自己打电话确认招聘者已收到履历表和自荐信并安排面试，语气肯定但要有礼貌。

示例如下：

I will connect with human resource department in two weeks, it is convenient to arrange the time to check my qualification.

PART 6 第 6 章 Electronic Commerce

6.1 Electronic Commerce

6.1.1 Text

Electronic commerce (E-Commerce) is doing business through electronic media (Fig. 6-1). It means using simple, fast and low-cost electronic communications to transact, without face-to-face meeting between the two parties of the transaction. Now, it is mainly done through Internet and Electronic Data Interchange (EDI). E-Commerce was first developed in the 1960s. With the wide use of computer, the maturity and the wide adoption of Internet, the permeation of credit cards, the establishment of secure transaction agreement and the support and promotion by governments, the development of E-Commerce is becoming prosperous, with people starting to use electronic means as the media of doing business. [1]

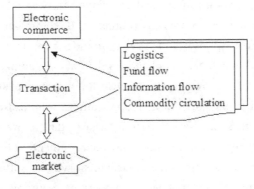

Fig. 6-1 Electronic commerce conceptual model

As computer network facilitates information exchange in a speedy and inexpensive way, Internet now penetrates into almost every corner of the world. Small and medium sized enterprises can forge global relationships with their trading partners everywhere in the world. High-speed network makes geographical distance insignificant. Businesses can sell goods to customers outside traditional markets, explore new markets and realize business opportunities more easily. Businesses can maintain their competitive advantage by establishing close contact with their customers and consumers at anytime through Internet by providing the latest information on products and services round the clock. [2] Internet provides companies with many markets in the cyber-world and numerous chances for product promotion. Besides, relationships with buyers can also be enhanced. By the use of multimedia capabilities, corporate image, product and service brand names can be established effectively through the Internet. Detailed and accurate sales data can help to reduce stock level and thus the operating cost. Detailed client information such as mode of consumption, personal preferences and purchasing power, etc. can help businesses to set their marketing strategies more effectively.

Before setting up E-commerce, you must make sure these questions: what is your objective of setting up E-commerce? To what extent do you want your company to adopt E-commerce? How much you have to invest? To setup E-commerce, you also have to consider the hardware and software in your company, and the technology required for setting up E-commerce.

1. E-commerce Model

The characteristics of E-commerce that distinguish it from traditional commerce are the means used for conveying information and the methods used for processing it. To effect changes in the way information is conveyed and processed, two support services are clearly needed: communications and data management. In addition, for realistic use of E-commerce applications in an operational setting, security is also essential.

The model for E-commerce must include communications, data management, and security. Of course, the application must also be part of the model. So a model for E-commerce comprised of five major components: a user interface, an application, communications, data management, and security. Notably, the model includes a user interface element which has seemingly not been included in any of the above discussion. This is because the user interface and the application are typically coupled so tightly that they often exist as a single entity which cannot be separated.

So creating a ubiquitous E-commerce environment will require deploying a network capable of providing connectivity to a large user and service provider community. Hardware, software, and security issues must be addressed. New applications that take advantage of networks and improved computer performance will be required.[3] Networking software that utilizes the increased bandwidth will be needed to support the new applications. Applications beyond those currently envisioned will evolve, and sophisticated user interfaces that allow users to conveniently take advantage of the services provided by this new environment will emerge. One of the key components for developing an E-commerce environment will be software that can support local applications and interoperate in a heterogeneous networking environment; this will require the use of open systems based on national and international standards.

2. Electronic Payment System

An Electronic Payment System (EPS) is needed for making payments towards purchases of information, goods and services provided through the Internet. Examples are sale of books, music

CDs, software, personal computers, Website maintenance services, research reports etc. It is also convenient form of payment for merchandise and services provided outside the Internet. EPS helps to automate sales activities, extends the potential number of customers. Electronic payments offer a less costly and more efficient alternative.

(1) Smart cards

A smart card is a small electronic device about the size of a credit card that contains electronic memory, and an embedded integrated circuit (IC). Smart cards containing an IC are sometimes called integrated circuit cards. To use a smart card, either to pull information from it or add data to it, you need a smart card reader, a small device into which you insert the smart card.[4] One of the benefits of the smart card technology is that the consumer can bring the ATM to his computer. Cash could be easily transferred over the computer from the bank account to the digital tokens in the smart card. The user can use this digital money for on-line Internet payment to vendors. This chip to chip money transfers has state of the art security mechanisms of private and public key encryption decryption algorithms.

(2) Credit cards

Credit cards, (and debit cards that share their networks) are the preferred method of payment among online consumers. Reasons for credit cards popularity as a payment method: Easy to use, no technology hurdles, and almost everyone has one; Consumers trust card companies to conduct secure transactions; Most merchants accept them. Parties involved in the credit card transactions are customer, customer's credit card Issuer (Issuing Bank), merchant, merchant's bank, and company processing credit card transactions over the Internet.

(3) Electronic cash

One needs to buy occasionally small value items like single newspaper. For online procurement of such items credit card is not preferred due to high transaction costs involved. Systems such as E-cash allow the customer to deposit cash into a bank account and receive an encoded digital string for each monetary unit, which is transferred into user's hard disk.[5] The customer can then transfer the cash to vendors on the Internet who also have supporting infrastructure. The vendor can then redeem this digital money into actual cash by depositing that in the bank.

3. Influence that E-commerce May Have

Compared with traditional commerce, E-commerce has superiorities as follows.

(1) Extensive coverage

A network system combining Internet, Intranet (local area network inside enterprises) and Extranet (networks outside enterprises) enables buyers, sellers, manufacturers and their partners to contact with each other and conveniently transmit commercial intelligence and documents worldwide.

(2) Complete functions

In E-commerce, users of different types and on different tiers can realize different targets in trade, for example, releasing commercial intelligence, on-line negotiation, electronic payment, establishment of virtual commercial marketplace and on-line banking, etc.

(3) Convenience and flexibility in use

Based on Internet, E-commerce is free from restriction by specialized protocol for data exchange. Transactions can be conducted conveniently on computer screen, by using any type of PCs, at any place around the world.

(4) Low cost

Use of E-commerce can cut down costs for hiring employees, maintaining warehouse and storefront, expense for international travel and postage to a great extent. The cost for using Internet is very low.

E-commerce will have substantial influence on social economy.

(1) E-commerce will change the way people used to take in commercial activities

Through networks, people can enter virtual stores browse around, select what they are interested in, and enjoy various on-line services. On the other hand, merchants can contact with consumers through networks, decide on buying in goods (categories and quantities) and perform settlement of accounts. Government agencies can perform electronic tendering and pursue government purchase through networks.

(2) The core of E-commerce is people

It is a social system. On-line shopping changes the way of people's daily life and fully embodies autonomy of consumers in trade.

(3) E-commerce changes the way enterprises produce their production

Through networks, manufactures know market demand directly and make arrangement of production, in accordance with consumers' need.

(4) E-commerce dramatically raises efficiency of trade

Intermediate links can be cut down, costs for sales will be reduced to minimum. Production can be arranged in "small batches plus diverse varieties", and "zero stock" will be reality.

(5) E-commerce calls for reformation of banking services

New concepts like on-line bank, on-line cash card and credit card, on-line settlement of accounts, electronic invoice, electronic "cash"—consumers will no longer use the real cash when shopping—will become reality.

(6) E-commerce will change government behavior

E-commerce is called "on-line government". It is an on-line administration.

Key Words

agency	代理机构，中介
competitive	竞争的
connectivity	连接
consumption	消费
deploy	展开，配置
dramatically	突然地，引人注意地
enterprise	企业
heterogeneous	异种的，不同的
hire	租用，租借

invest	投资
maturity	成熟，偿还期，到期
monetary	货币的
negotiation	谈判，磋商
payment	付款，支付
penetrate	穿透
permeation	渗透
pursue	实行，实现
tender	温柔的，审慎的
transact	办理，处理
ubiquitous	普遍存在的
warehouse	仓库
zero stock	零库存

Notes

[1] With the wide use of computer, the maturity and the wide adoption of Internet, the permeation of credit cards, the establishment of secure transaction agreement and the support and promotion by governments, the development of E-Commerce is becoming prosperous, with people starting to use electronic means as the media of doing business.

译文：随着计算机的广泛应用，互联网的日趋成熟和广泛利用，信用卡的普及，政府对安全交易协定的支持和促进，电子商务正变得繁荣起来，人们已开始利用电子媒介做生意。

本句的主句是"the development of E-Commerce is becoming prosperous"，主句之前的是伴随状语；"maturity"的意思是"成熟，完备，（票据）到期"，"permeation"原意是"渗入，透过"，这里指信用卡在社会中的普及。

[2] Businesses can maintain their competitive advantage by establishing close contact with their customers and consumers at anytime through Internet by providing the latest information on products and services round the clock.

译文：通过在互联网上全天候地提供产品及服务的最新信息，商家可以与客户和消费者随时建立紧密联系来确保他们的竞争优势。

本句的"by establishing…"是状语"latest information"的意思是"最新信息"，"round the clock"的意思是"24小时"。

[3] New applications that take advantage of networks and improved computer performance will be required.

译文：需要那些利用网络优势，提升计算机性能的新的应用。

本句由"that"引导定语从句，修饰主语。

[4] To use a smart card, either to pull information from it or add data to it, you need a smart card reader, a small device into which you insert the smart card.

译文：使用智能卡，无论是从中提取信息还是向卡中加入数据，都需要一台智能卡读卡机，一台将智能卡插入其中的小设备。

本句的"To use a smart card"是目的状语，"either to pull information from it or add data to it"是让步状语，"a small device into which you insert the smart card"是非限制性定语，修饰"smart card reader"。

[5] Systems such as E-cash allow the customer to deposit cash into a bank account and receive an encoded digital string for each monetary unit, which is transferred into user's hard disk.

译文：电子现金系统，例如 E-cash，允许顾客将现金存放在一个银行账户中，则顾客将收到一个代表一个货币单位的编码数字串，该数字串被传送到用户的硬盘之中。

本句的"such as E-cash"作主语的同位语，"which is transferred into user's hard disk"是非限制性定语从句。

6.1.2 Exercises

1. Translate the following phrases into English or Chinese

（1）bank account
（2）digital money
（3）face-to-face
（4）trading partner
（5）geographical distance
（6）电子数据交换
（7）信用卡
（8）电子媒介
（9）电子支付系统
（10）集成电路

2. Identify the following to be True or False according to the text

（1）As computer network facilitates information exchange in a speedy and expensive way.
（2）Electronic payments offer a more costly and more efficient alternative.
（3）Credit cards are the preferred method of payment among traditional consumers.
（4）Smart cards containing an IC are sometimes called Integrated Circuit Cards.
（5）Internet now penetrates into almost every corner of the world.
（6）In electronic commerce, users of different types and on different tiers can realize different targets in trade.

3. Reading Comprehension

（1）E-Commerce was first developed in the_____.
A. 1950s
B. 1960s
C. 1970s
D. 1980s

（2）By the use of multimedia capabilities, corporate image, product and service brand names can be established effectively through_____.
A. the computer
B. the Internet
C. the tools
D. the system

(3) The characteristics of electronic commerce that distinguish it from _____ are the means used for conveying information and the methods used for processing it.
   A. traditional commerce
   B. modern commerce
   C. newer commerce
   D. computer-based commerce

(4) The user can use this digital money for on-line Internet payment to _____.
   A. programmer
   B. machine
   C. system
   D. vendors

(5) _____ will change the way people used to take in commercial activities.
   A. Traditional commerce
   B. Commerce
   C. Electronic commerce
   D. Original commerce

6.1.3 Reading Material

Business Networking Technology

Electronic commerce is a system that includes not only those transactions that center on buying and selling goods and services to directly generate revenue, but also those transactions that support revenue generation, such as generating demand for those goods and services, offering sales support and customer service, or facilitating communications between business partners. Electronic commerce builds on the advantages and structures of traditional commerce by adding the flexibilities offered by electronic networks.

Electronic commerce enables new forms of business, as well as new ways of doing business. Amazon.com, for example, is a bookseller based in Seattle, Washington. The company has not physical stores, sells all their books via the Internet, and coordinates deliveries directly with the publishers. Because all of their products (commercial software packages) are electronic, and can be stored on the same computers that they use for processing orders and serving the web, their inventory is totally digital.

Information technology makes new business solutions possible. This might mean new or improved products and services (e. g. automobile and navigation), additional sales channels (e. g. Internet banking), more efficient forms of procurement (e. g. global procurement by means of electronic markets), new ways in which suppliers and customers can cooperate (e. g. collaborative planning), new services (e. g. virtual communities), more effective management (e. g. through the automatic measurement of key performance indicators) or new information services (e. g. product catalogs).

Imagine a scenario in which every employee, every customer, every business partner, every appliance and every computer has immediate access to each other's data at any time.

Network makes it possible to offer all products and services for a customer process on a coordinated basis and from a single source. It is not the customer but the supplier who is the specialist in the process car ownership. The car dealer takes on the task of helping the customer with car selection, obtaining the test reports, financing, navigation, resale, etc. This applies equally to the stages upstream of the car dealer, e.g. completeness of a car journal's services, perhaps including online research in earlier issues or access to sources. The credit scoring institute provides not only the customer's credit rating but also concrete information on delays in payment or loan insurance, the tire dealer supplies not only the tires but also the CAD data (computer-aided design) and test reports as well as batch quality data, etc.

Seen by a business point of view the numerous IT developments are responsible for seven fundamental trends in business transformation.

- Enterprise resource planning (ERP), i. e. the operational execution of business, runs almost imperceptibly in the background. Integrated applications for administration as well as for product development and technology make it possible to concentrate on business rather than on administration.
- Knowledge management supplies each task within a process with the necessary knowledge about customers, competitors, products, etc. and above all about the process itself.
- Smart appliances take information processing to the point of action. Traffic information is supplied via the satellite navigation system (GPS) to the motorist, point of sale information from the cash register to the product manufacturer and machine faults via sensors to the service engineer.
- Business networking makes collaboration between two companies so simple that they appear to be one enterprise. Information on sales of the end product is immediately available to all the companies in the supply chain.
- Many sub-processes which companies still operate individually at the moment will be available from the net as electronic services. One example could be customer profiling. In addition to the supplier, a third party online-database provider and the customer him or herself can take over the responsibility of his or her profile and offer it via an electronic service.
- Companies will not simply be selling products or services but will be supporting entire customer processes. Transport businesses will take on the logistics process, doctors will support the whole therapy process and insurance companies will handle the claim processing instead of the customer.
- Corporate management will no longer merely focus on financial results but also on factors contributing to these results. Financial management will become value management which keeps an eye on key performance indicators for the success of the business.

6.2 Web Navigation

6.2.1 Text

Web navigation design is about linking. It is about determining importance and relevance of the pages and content on your site. This requires judgment in establishing meaningful relationships between pages of information. Together, the elements of navigation determine not only if you can find the information people are looking for, but also how you experience that information.[1]

A common goal in navigation design is to create effortless interaction with information. Navigation should be "invisible" to the user. Measuring its effectiveness is therefore problematic. It's difficult to demonstrate the value of something that is at its best when you don't notice it. At the same time, navigation is deceptively complex. The thousands of pages you have provided access to, the numerous relationships you have established between different pieces of content, and the smooth interaction detailed in countless flowcharts all get reduced to a handful of links. The navigation on any one page is just a small portion of a larger system, which is sometimes hard to grasp.

When navigating your site, visitors should be informed about what's going on. The navigation system you design gives them cues as to how to navigate through the site. Text and labels are the primary way people will know which option is which or what the title of the current page is. But beyond this, feedback in navigation can be considered in two ways: with rollovers before selecting a navigation option, and by showing location after transitioning to a new page.

Showing the position of a page within a site by highlighting its category in the navigation helps orient visitors to their location. On large information-rich sites, this is valuable to orientation. Showing location also helps if someone gets interrupted and needs to resume their session later.

Before beginning to design web navigation, you should take time to investigate how people look for information. The more you understand, the easier it is to structure your thoughts around solving design problems.[2] Web navigation is defined three ways:

(1) The theory and practice of how people move from page to page on the web.

(2) The process of goal-directed seeking and locating hyperlinked information: browsing the web.

(3) All of the links, labels, and other elements that provide access to pages and help people orient themselves while interacting with a given web site.

Links are text or graphics in one page that connect to another page or a different location within a single page. They allow for the associative leap from one idea to the next. If you are reading a story about China's foreign policy you can then jump to a page with detailed information about the demographics of the country, thanks to a link.[3] But, links do more than just associate one page with another. They also show importance. Links show relevance.

Web navigation plays a major role in shaping our experiences on the web. It provides access to information in a way that enhances understanding, reflects brand, and lends to overall credibility of a site. And ultimately, web navigation and the ability to find information have a financial impact for stakeholders.

Navigation design is a task that is not merely limited to choosing a row of buttons. It is much broader, and, at the same time, more subtle than that. The navigation designer coordinates user goals with business goals. This requires an understanding of each, as well as a deep knowledge of information organization, page layout, and design presentation.

Overall, the various elements come together to create system of navigation. Though visitors might perceive this system as whole, we can dissect its individual components. For instance, the tabs at the top center of the page (starting with Home) are referred to as the main navigation.

Is this a good navigation system? The answer is ultimately relative. You must consider a range of factors, from business goals to user goals. Still, there are common principles of good navigation that we can use to evaluate the quality of navigation. For instance, navigation on the BBC page is balanced, consistent, and provides a clear indication of where you are. Overall, it is appropriate for this type of site and probably gets people to where they want to go—the most important factor for judging success in navigation design.

The following sections outline some of the more important qualities of successful navigation. These are not rules. They don't prescribe how to design navigation. But understanding them can guide your thought process while designing. Overall, these aspects predict the effectiveness of a navigation system.

- Balance.
- Ease of learning.
- Consistency (and inconsistency).
- Feedback.
- Efficiency.
- Clear labels.
- Visual clarity.
- Appropriateness for the site type.
- Aligning with user goals.

On web sites, breadth and depth is a function of both the information structure and the navigation. Directories, for instance, often show options to two levels of the hierarchy on one page, thereby reducing the number of clicks to get to a second-level page. Dynamic menus have a similar effect in that they can access deeper pages in a site directly from a top-level page.

Generally, broader structures work better than deeper ones. The effect it takes to continually choose categories across many levels of a deep structure outweighs the effect to scan many items in a broad navigation. The eye is much faster than a mouse click (and page load). Although users tend to become (and possibly lost) quicker in deeper structures, don't swing too far toward breadth. Showing all links at all times can be overwhelming and make choices harder. Visitors may just take the first option that appears to fit their need, or simply give up.

The paths to information should be efficient. Strive to create navigational links, tables, and icons that are easy to see and easy to click. For instance, dynamic menus that require hand-eye coordination just to reach the options will slow users down. Interacting with the link, buttons, tables, and menus you create should require minimum effort.

A common tactic in navigation design is to cluster options, grouping like items together to provide layers of focus. Users then don't have to scan every link on the page; they can look at component headings. From there they can focus on the links in that area. It is a two-stage scanning process: first find the right group, then, zoom in on the individual links.

People don't expect that they will need to learn how to use a web site. There is no training period or expectation of having to study a manual or set of instructions first. On the open web, the duration of time spent on a single page is typically measured in seconds. The intent and function of the navigation must be immediately clear, particularly for information-rich sites with business goals, but also for any type of web navigation.[4]

Be "consistent" is a primary guideline of interface design. In terms of navigation, this usually refers to the mechanisms and links that appear in a steady location on the page, behave predictably, have standardized labels, and look the same across the site. Generally, this is good approach and something you should strive for. But keep in mind that consistency does not equal uniformity; inconsistency is just as important in navigation design. Things that work in different ways should also differ in appearance. The real rule when people say to "be consistent" is: "balance consistency with inconsistency". Some inconsistency is critical to navigation. Varying the position, color, labels, or general layout of a mechanism creates a sense of progress through a site.[5]

Key Words

align	匹配
consistency	一致，相容
credibility	可靠性，确实性
deceptively	靠不住地，虚伪地
dissect	仔细分析，剖开
effortless	不费力的，容易的
hierarchy	分层，层次，分类等级
inconsistency	不一致，不一贯，不协调
leap	跳越，跳过
overwhelming	势不可当的，压倒的
perceive	领悟，领会
prescribe	命令，指挥
problematic	疑难的，未定的
progress	进展，进行
relevance	有关系，适当
resume	恢复，重新开始
stakeholder	利益共享者
uniformity	一样，一致，一式

Notes

[1] Together, the elements of navigation determine not only if you can find the information people are looking for, but also how you experience that information.

译文：同时，通过导航元素不仅能够找到人们正在找的信息，而且能够以不同的方式体验这些信息。

本句用"not only...but also..."结构表示两种情况。

[2] The more you understand, the easier it is to structure your thoughts around solving design problems.

译文：你考虑得越多，在解决设计问题时，就越容易构建你的想法。

本句用"the more...the easier..."比较级结构。

[3] If you are reading a story about China's foreign policy you can then jump to a page with detailed information about the demographics of the country, thanks to a link.

译文：例如你正在读关于中国的外交政策的信息，你能跳转到关于国家人口统计详细信息的页面，这就是链接的功劳。

本句由"If"引导条件状语从句。

[4] The intent and function of the navigation must be immediately clear, particularly for information-rich sites with business goals, but also for any type of web navigation.

译文：导航的意向和功能一定是迅捷且清晰的，特别是那些有企业目标的资源丰富的站点，这也对任何类型的网页导航有用。

本句的"particularly for..."是状语。

[5] Varying the position, color, labels, or general layout of a mechanism creates a sense of progress through a site.

译文：通过在网站中改变位置、颜色、标签或者通用布局能创造一种进展感觉。

本句由分词短语"Varying the position, color, labels, or general layout of a mechanism"作主语。

6.2.2 Exercises

1. Translate the following phrases into English or Chinese

（1）web navigation

（2）page layout

（3）design presentation

（4）top-level page

（5）navigational link

（6）信息组织

（7）两阶段扫描

（8）导航系统

（9）导航设计

（10）动态菜单

2. Identify the following to be True or False according to the text

（1）Showing the position of a page within a site by highlighting its category in the navigation helps orient visitors to their location.

（2）Before beginning to design web navigation, you should take time to investigate how people look for information.

（3）A common tactic in navigation design is to cluster options, grouping like items together to provide layers of focus.

（4）When navigating your site, visitors should not be informed about what's going on.

（5）Navigation should be "visible" to the user.

（6）Links are text or graphics in one page that connect to another page or a different location within a single page.

3. Reading Comprehension

（1）For instance, _____ on the BBC page is balanced, consistent, and provides a clear indication of where you are.

A. data

B. graph

C. information

D. navigation

（2）The navigation designer coordinates user goals with_____.

A. business goals

B. leader goals

C. technique goals

D. their own goals

（3）The navigation on any _____ is just a small portion of a larger system, which is sometimes hard to grasp.

A. one page

B. pages

C. search

D. process

（4）_____ plays a major role in shaping our experiences on the web.

A. Page layout

B. Web navigation

C. Program

D. Database

（5）On the web, the duration of time spent on a page is typically measured in_____.

A. minutes

B. hours

C. seconds

D. milliseconds

6.2.3　Reading Material

Social Networking

Social networking is the grouping of individuals into specific groups, like small rural communities or a neighborhood subdivision. Although social networking is possible in person, especially in the workplace, universities, and high schools, it is most popular online.

This is because unlike most high schools, colleges, or workplaces, the Internet is filled with millions of individuals who are looking to meet other people, to gather and share first-hand information and experiences about cooking, golfing, gardening, developing friendships professional alliances, finding employment, business-to-business marketing and even groups sharing information about baking cookies to the thrive movement. The topics and interests are as varied and rich as the story of our universe.

When it comes to online social networking, Websites are commonly used. These Websites are known as social sites. Social networking Websites function like an online community of Internet users. Depending on the Website in question, many of these online community members share common interests in hobbies, religion, politics and alternative lifestyles. Once you are granted access to a social networking Website you can begin to socialize. This socialization may include reading the profile pages of other members and possibly even contacting them.

The friends that you can make are just one of the many benefits to social networking online. Another one of those benefits includes diversity because the Internet gives individuals from all around the world access to social networking sites. This means that although you are in the United States, you can develop an online friendship with someone in Denmark or India. Not only will you make new friends, but you just might learn a thing or two about new cultures or new languages and learning is always a good thing.

As mentioned, social networking often involves grouping specific individuals or organizations together. While there are a number of social networking Websites that focus on particular interests, there are others that do not. The websites without a main focus are often referred to as "traditional" social networking Websites and usually have open memberships. This means that anyone can become a member, no matter what their hobbies, beliefs, or views are. However, once you are inside this online community, you can begin to create your own network of friends and eliminate members that do not share common interests or goals.

As we know, there are dangers associated with social networking including data theft and viruses, which are on the rise. The most prevalent danger though often involves online predators or individuals who claim to be someone that they are not. Although danger does exist with networking online, it also exists in the real world too. Just like you are advised when meeting strangers at clubs and bars, schools, or work—you are also advised to proceed with caution online.

By being aware of your cyber-surroundings and who you are talking to, you should be able to safely enjoy social networking online. It will take many phone conversations to get to know someone, but you really won't be able to make a clear judgment until you can meet each other in person.

Once you are well informed and comfortable with your findings, you can begin your search from hundreds of networking communities to join. This can easily be done by performing a standard Internet search. Your search will likely return a number of results, including MySpace, FriendWise, FriendFinder, Yahoo! 360, Facebook, Orkut, and Classmates.

6.3 Mobile Commerce

6.3.1 Text

With the introduction of the World Wide Web, electronic commerce has revolutionized traditional commerce and boosted sales and exchanges of merchandise and information. Recently, the emergence of wireless and mobile networks has made possible the admission of electronic commerce to a new application and research subject: mobile commerce, which is defined as any transaction with a monetary value that is conducted via a mobile telecommunications network.[1] A somewhat looser approach would be to characterize mobile commerce as the emerging set of applications and services people can access from their Internet-enabled mobile devices. Mobile commerce is an effective and convenient way to deliver electronic commerce to consumers from anywhere and at anytime. Realizing the advantages to be gained from mobile commerce, many major companies have begun to offer mobile commerce options for their customers.[2]

It is a short number of steps from owning a smart phone or tablet, to searching for products and services, browsing, and then purchasing (Fig. 6-2). The resulting mobile commerce is growing at over 50% a year, significantly faster than desktop e-commerce at 12% a year. The high rate of growth for mobile commerce will not, of course, continue forever; but analysts estimate that by 2017, mobile commerce will account for 18% of all e-commerce.

Fig. 6-2 Mobile commerce

A study of the top 400 mobile firms by sales indicates that 73% of mobile commerce is for retail goods, 25% for travel, and 2% for ticket sales. Increasingly, consumers are using their mobile devices to search for people, places, and things—like restaurants and deals on products they saw in a retail store. The rapid switch of consumers from desktop platforms to mobile devices is driving a surge in mobile marketing expenditure.

Compared to an electronic commerce system, a mobile commerce system is much more complicated because components related to mobile computing have to be included.[3] The following outline gives a brief description of a typical procedure that is initiated by a request submitted by a mobile user.

- Mobile commerce applications: A content provider implements an application by providing two sets of programs: client-side programs, such as a user interface on a micro-browser, and server-side programs, such as database accesses and updating.
- Mobile stations: Mobile stations present user interfaces to the end users, who specify their requests on the interfaces. The mobile stations then relay user requests to the other components and display the processing results later using the interfaces.
- Mobile middleware: The major purpose of mobile middleware is to seamlessly and transparently map Internet contents to mobile stations that support a wide variety of operating systems, markup languages, micro-browsers, and protocols. Most mobile middleware also encrypts the communication in order to provide some level of security for transactions.
- Wireless networks: Mobile commerce is possible mainly because of the availability of wireless networks. User requests are delivered to either the closest wireless access point (in a wireless local area network environment) or a base station (in a cellular network environment).
- Wired networks: This component is optional for a mobile commerce system.
- Host computers: This component is similar to the one used in electronic commerce. User requests are generally acted upon in this component.

Without trust and security, there is no mobile commerce period. How could a content provider or payment provider hope to attract customers if it cannot give them a sense of security as they connect to paying services or make purchases from their mobile devices? Consumers need to feel comfortable that they will not be charged for services they have not used, that their payment details will not find their way into the wrong hands, and that there are adequate mechanisms in place to help resolve possible disputes.

As we review different aspects of mobile security, it is important to keep in mind that security always requires an overall approach. A system is only as secure as its weakest component, and securing networking transmission is only one part of the equation. Technically speaking, there are a number of different dimensions to network security, each corresponding to a different class of threat or vulnerability. Protecting against one is no guarantee that you will not be vulnerable to another.

The applications of mobile commerce are widespread. The following lists some of the major mobile commerce applications along with details of each.

- Commerce

Commerce is the exchange or buying and selling of commodities on a large scale involving transportation from place to place. It is boosted by the convenience and ubiquity conveyed by mobile commerce technology. There are many examples showing how mobile commerce helps commerce. For example, consumers can now pay for the products in a vending machine or a parking fee by using their cellular phones; mobile users can check their bank accounts and perform account balance transfers without needing to go a bank or access an ATM; etc.

- Education

Many schools and colleges are facing problems due to a shortage of computer lab space, separation of classrooms and labs, and the difficulty of remodeling old classrooms for wired networks. To solve these problems, wireless LANs are often used to hook PCs or mobile handheld devices to the Internet and other systems.[4] As a result, students are able to access many of the required resources without needing to visit the labs.

- Entertainment

Entertainment has always played a crucial role in Internet applications and is probably the most popular application for the younger generation. Mobile commerce makes it possible to download games / images / music / video files at anytime and anywhere, and it also makes online games and gambling much easier to access.

- Health Care

The cost of health care is high and mobile commerce can help to reduce it. By using the technology of mobile commerce, physicians and nurses can remotely access and update patient records immediately, a function that has often incurred a considerable delay in the past. This improves efficiency and productivity, reduces administrative overheads, and enhances overall service quality.

- Inventory Tracking and Dispatching

Just-in-time delivery is critical for the success of today's businesses. Mobile commerce allows a business to keep track of its mobile inventory and make time definite deliveries, thus improving customer service, reducing inventory, and enhancing a company's competitive edge.[5] Most major delivery services, such as UPS and FedEx, have already applied these technologies to their business operations worldwide.

- Traffic

Traffic control is usually a major headache for many metropolitan areas. Using the technology of mobile commerce can easily improve traffic in many ways. For example, it is expected that a mobile handheld device will have the capabilities of a GPS (Global Positioning System), e.g., determining the driver's position, giving directions, and advising on the current status of traffic in the area; a traffic control center could monitor and control the traffic according to the signals sent from mobile devices in vehicles.

- Travel and Ticketing

Travel expenses can be costly for a business. Mobile commerce could help reduce operational costs by providing mobile travel management services to business travelers.

Key Words

adequate	足够的，恰当的
administrative	管理的，行政的
analyst	分析家，化验员
boost	增加，促进，提高
cellular	蜂窝状，多孔的
comfortable	舒适的，处于轻松的
commodity	商品，日用品
considerable	相当大的，应考虑的

crucial	决定性的，关键性的
description	描述，形容
dispute	辩论，争论
expenditure	经费，费用，支出
handheld	手持，手握
merchandise	商品，货物，经商
metropolitan	大城市的，大都市的
middleware	中间件
monetary	货币的，金钱的
purchase	购买，采购
seamlessly	无空隙地，无停顿地
telecommunication	电信
threat	威胁，恐吓
transparently	显然地，易察觉地
trust	信任，相信
ubiquity	到处存在，普遍存在
vulnerability	脆弱性，致命的

Notes

[1] Recently, the emergence of wireless and mobile networks has made possible the admission of electronic commerce to a new application and research subject: mobile commerce, which is defined as any transaction with a monetary value that is conducted via a mobile telecommunications network.

译文：最近，无线网络和移动网络的出现使得电子商务进入了新的应用及科研主题：移动商务。移动商务可以定义为：通过移动的无线电通信网络进行的涉及货币价值的任何交易。

本句的"wireless and mobile networks"作主语的定语，"which is defined as…"作非限定性定语从句。

[2] Realizing the advantages to be gained from mobile commerce, many major companies have begun to offer mobile commerce options for their customers.

译文：很多大公司已经意识到移动商务带来的好处，开始为客户提供移动商务选择。

本句的分词结构"Realizing the advantages…"作状语。

[3] Compared to an electronic commerce system, a mobile commerce system is much more complicated because components related to mobile computing have to be included.

译文：与电子商务系统相比，移动商务系统要复杂得多，因为它必须包含与移动计算相关的成分。

本句的"Compared to an electronic commerce system"是条件状语，由"because"引导的是原因状语从句。

[4] To solve these problems, wireless LANs are often used to hook PCs or mobile handheld devices to the Internet and other systems.

译文：为了解决这些问题，无线局域网通常用来将计算机或移动手持设备连接到互联网或其他系统。

本句的"To solve these problems"作目的状语。

[5] Mobile commerce allows a business to keep track of its mobile inventory and make time definite deliveries, thus improving customer service, reducing inventory, and enhancing a company's competitive edge.

译文：移动商务让企业可以跟踪其移动库存并适时地投递，这样可以改善客户服务、减少库存并增强公司的竞争力。

本句的"to keep track of …"作宾语补足语，"thus improving customer…"是目的状语。

6.3.2 Exercises

1. Translate the following phrases into English or Chinese

（1）high rate of growth
（2）mobile marking
（3）content provider
（4）mobile inventory
（5）mobile station
（6）桌面电子商务
（7）银行账户
（8）移动商务
（9）零售物品
（10）智能卡

2. Identify the following to be True or False according to the text

（1）Entertainment has always played a crucial role in Internet applications and is probably the most popular application for the younger generation.

（2）A study of the top 400 mobile firms by sales indicates that 75% of mobile commerce is for retail goods.

（3）Commerce is the exchange or buying and selling of commodities on a large scale involving transportation from place to place.

（4）It is a short number of steps from owning a smart phone or tablet, to searching for products and services, browsing, and then purchasing.

（5）Compared to an electronic commerce system, a mobile commerce system is not much more complicated.

（6）Mobile commerce is possible mainly because of the availability of wireless networks.

3. Reading Comprehension

（1）The resulting mobile commerce is growing at over 50% a year, significantly faster than desktop e-commerce at _____ a year.

A. 12%
B. 15%
C. 18%
D. 20%

(2) A _____ implements an application by providing two sets of programs: client-side programs, such as a user interface on a micro-browser, and server-side programs.

A. network provider
B. content provider
C. service provider
D. interface provider

(3) Consumers need to feel _____ that they will not be charged for services they have not used.

A. afraid
B. comfortless
C. comfortable
D. sorry

(4) There are a number of different dimensions to _____, each corresponding to a different class of threat or vulnerability.

A. network
B. network security
C. electronic commerce
D. mobile commerce

(5) Mobile commerce is an effective and convenient way to deliver electronic commerce to from _____ anywhere and at anytime.

A. bank
B. sellers
C. someone
D. consumers

6.3.3 Reading Material

The Internet of Things

The Internet of Things builds on foundation of existing technologies, such as RFID, and is being enabled by the availability of low cost sensors, the drop in price of data storage, the development of "Big Data" analytics software that can work with trillions of pieces of data, as well implementation of IPv6, which will allow Internet addresses to be assigned to all of these new devices (Fig. 6-3).

Fig. 6-3　The Internet of Things

The Internet of Things, sometimes referred to as the Internet of Objects, will change everything—including ourselves. This may seem like a bold statement, but consider the impact the Internet already has had on education, communication, business, science, government, and humanity. Clearly, the Internet is one of the most important and powerful creations in all human history.

From a technical point of view, the Internet of Things is not the result of a single novel technology; instead, several complementary technical developments provide capabilities that taken together help to bridge the gap between the virtual and physical world. These capabilities include:

- Communication and cooperation: Objects have the ability to network with Internet resources or even with each other, to make use of data and services and update their state. Wireless technologies such as GSM and UMTS, Wi-Fi, Bluetooth, ZigBee and various other wireless networking standards, are of primary relevance here.
- Addressability: Within an Internet of Things, objects can be located and addressed via discovery, look-up or name services, and hence remotely interrogated or configured.
- Identification: Objects are uniquely identifiable. RFID, NFC (Near Field Communication) and optically readable bar codes are examples of technologies with which even passive objects which do not have built-in energy resources can be identified (with the aid of a "mediator" such as an RFID reader or mobile phone). Identification enables objects to be linked to information associated with the particular object and that can be retrieved from a server, provided the mediator is connected to the network.
- Sensing: Objects collect information about their surroundings with sensors, record it, forward it or react directly to it.
- Actuation: Objects contain actuators to manipulate their environment (for example by converting electrical signals into mechanical movement). Such actuators can be used to remotely control real-world processes via the Internet.
- Embedded information processing: Smart objects feature a processor or microcontroller, plus storage capacity. These resources can be used, for example, to process and interpret sensor information, or to give products a "memory" of how they have been used.
- Localization: Smart things are aware of their physical location, or can be located. GPS or the mobile phone network is suitable technologies to achieve this, as well as ultrasound time measurements, UWB (Ultra-Wide Band), radio beacons (e.g. neighboring WLAN base stations or RFID readers with known coordinates) and optical technologies.
- User interfaces: Smart objects can communicate with people in an appropriate manner (either directly or indirectly, for example via a smart phone). Innovative interaction paradigms are relevant here, such as tangible user interfaces, flexible polymer-based displays and voice, image or gesture recognition methods.

Most specific applications only need a subset of these capabilities, particularly since implementing all of them is often expensive and requires significant technical effort. Logistics applications, for example, are currently concentrating on the approximate localization (i.e. the position of the last read point) and relatively low-cost identification of objects using RFID or bar codes. Sensor data (e.g. to monitor cool chains) or embedded processors are limited to those logistics applications where such information is essential such as the temperature-controlled transport of vaccines.

As cool as the Internet of Things sounds, there is a downside. Issues such as privacy, reliability, and control of data still have to be worked out.

6.4 Electronic Payment and Logistics

6.4.1 Text

Electronic Payment

Payment is a very important step in the E-Commerce procedure. The Internet payment means conducting fund transfer, money receiving and disbursing electronically. Internet payment, sometimes called electronic payment, is developed for the demand of E-Commerce applications. There are many kinds of digital money, or electronic token, which mainly could be classed into two categories: real-time payment mechanism, such as digital cash, electronic wallet and smart card, and after time payment mechanism, such as electronic check (net check) or electronic credit card (Web credit card).

The advanced encryption and authentication system can make digital money be used more safely and privately than paper money, and at the same time, keep the feature of "anonymity of payer" of paper money. The digital cash system is constructed based on digital signature and cryptograph techniques. This system mostly takes use of the public/private key pairs methods.

First, the client deposits money in the bank, and opens an account. The bank will give the client the digital cash software used for client-side and a bank's public key. The bank uses sever-side digital cash software and private key to encode the message. The client can decode such message by using client-side digital cash software and the bank's public key. When client needs some digital money to pay someone, he initiates that sum of digital money he needs by using the digital cash software, then, he encodes it and transfers the message to someone.[1] Someone encodes this message, his own account message and depositing orders and sends them to the bank. The bank decodes these messages with its private key and confirms the authenticity of both the client and someone. Then transfer the money from client's account to someone's account.

Since the digital cash is only a digital number, so it needs to be verified as authentic. Banks must have the ability to track whether the e-cash is duplicated or reused, without linking the personal shopping behavior to the person who bought that cash from the bank, to keep the feature of "anonymity of payer" of real money.[2] In the whole checking process, the seller can't capture any private data of his client. He can verify the digital cash sent by the client with the bank's public key.

Another mechanism for Internet payment is e-check. Its former is the financial EDI system used in Value Added Networks. Some electronic check software can be packed in smart card. Thus, clients can bring it very conveniently and can write checks at any computer site. Some electronic-check software can provide e-wallets for their customers, which also enable the customers to write checks at anywhere even they have no bank account.

An electronic bank is not a real bank, it is a virtual bank, in which, the bankers, software designers, and hardware producers and ISPs work together. Each of them fulfils its own work in its profession and coordinates perfectly with each other. All the services are conducted on the bank server. The bank interacts with its clients through website. The clients may use a portable smart card and conduct their bank businesses anywhere at any time, such as pay a bill, buy some digital money, fill out an electronic bank and so on.

These new types of bank server software, which adopt up the modern information technologies, can provide convenient, real-time, secure and reliable bank services.[3] They can also provide other relevant services, such as stock quotation, tax and polices consultation, exchange and interest rates information. The more convenient the software are, the more users there may be, the more profits the company will get and better services they provide, thus, the banks and the clients are in win/win relationship, clients get cheaper and convenient services. Banks greatly reduce the operating cost. For on, Internet, the cost of running a website is much lower than building a bank branch, and on Internet, the cost for serving one person is the same as that for serving 10 thousand persons. With the rapid development of electronic banks, software developing companies and ISPs will get more and more businesses, they are all in win/win relationship with electronic banks.

Logistics

As goods move, so must information. To move the right goods to the right place at the right time in the right condition with the right documents, the answers to all the "right" questions must be known. The essence of information systems in logistics is the conversion of accurate data into useful information. Inaccurate data and poor information disrupt logistics activities. Of cource, even with accurate data and good information, someone must act on it.

Peak logistics efficiency and effectiveness demand a superior integrated logistics information system (ILIS). Without ready access to accurate information, integrated logistics operations lose both efficiency and effectiveness. Integrated logistics will not sustain a strategic, competitive edge. Priority applications of ILIS are inventory status, tracing and expediting, pickup and delivery, order convenience, order accuracy, balancing inbound and outbound traffic opportunities, and order processing. The quality of the information flowing through the ILIS is of utmost importance. Three concerns stand out on quality information: (1) getting the right information, (2) keeping the information accurate, and (3) communicating the information effectively.

An integrated logistics information system can be defined as: the involvement of people, equipment, and procedures required to gather, sort, analyze, evaluate, and then distribute needed information to the appropriate decision-makers in a timely and accurate manner so they can make quality logistics decisions.

An ILIS gathers information from all possible sources to assist the integrated logistics manager in making decisions. It also interfaces with marketing, financial, and manufacturing information systems. All of this information is then funneled to top level management to help formulate strategic decisions.

The ILIS has four primary components: the order processing system, research and intelligence system, decision support system, and reports and outputs system. Together, these four subsystems should provide the integrated logistics manager with timely and accurate information on which to base decisions. These subsystems interface with the integrated logistics managerial functions and the integrated logistics management environment. Before information is developed, information needs must be determined. Likewise, once the information is generated based on a needs assessment, it is then distributed to the integrated logistics manager.

The Order Processing System is without a doubt the most important subsystem. Order processing is the set of activities necessary to make the correct goods ready for shipment to the customer—right up to the point where warehousing assembles the order. Processing the order includes checking customer credit, crediting a sales representative's account, ensure product availability, and preparing the necessary shipping documents. The seller should be able to control order cycle activity. Order processing time has been shortened largely through computer applications.

Research and Intelligence System (RIS) continually scans and monitors the environment, observing and drawing conclusions about the events that affect integrated logistics operations.[4] The RIS scans and monitors the intra-firm environment, the external environment, and the inter-firm environment. The external environment includes those events taking place outside the firm and normally out of the firm's control. The inter-firm environment includes elements in the external environment that directly affect the firm and over which the firm does exercise some control, such as the channel of distribution.[5] The intra-firm environment includes the internal work of the elements that are controlled by the firm.

Decision support systems (DSS) are computer-based and provide solutions to complex integrated logistics problems using analytical modeling. The heart of any DSS is a comprehensive database containing the information that integrated logistics managers can use to make decisions.

The final subsystem of ILIS is the reports and outputs system. Normal reports are used for planning, operating, and controlling integrated logistics. Planning output includes sales trends, economic forecasts, and other marketplace information. Operating reports are used in inventory control, transportation scheduling and routing, purchasing, and production scheduling. Control reports are used analyze expenses, budgets, and performance.

Key Words

anonymity	匿名，匿名者
authentic	可信的，可靠的
authentication	认证，证明
budget	预算，预算拨款
comprehensive	有理解力的，悟性好的
consultation	商议，磋商
deposit	保证金，储蓄
disburse	支出，付出
evaluate	评价，评估
funneled	漏斗状的，倾销
inbound	入境的，到达的

initiate	开始，发起，开辟
involvement	参与，加入
logistics	物流，后勤
outbound	出境的
performance	表现，表演
portable	轻便的，手提的
priority	优先，优先权
profession	职业，专业
profit	利润，收益
quotation	行情，行市
representative	有代表性的，典型的
reused	再生的，重用的
strategic	战略性的，战略上的
sustain	维持，支撑
utmost	极度的，最大的

Notes

[1] When client needs some digital money to pay someone, he initiates that sum of digital money he needs by using the digital cash software, then, he encodes it and transfers the message to someone.

译文：当客户需要一些数字现金支付某人时，就用数字现金软件生成他所需要的数字现金，给它加密并传给某人。

本句的"When…"是时间状语，"he initiates…"和"he encodes it and transfers the…"是并列句。

[2] Banks must have the ability to track whether the e-cash is duplicated or reused, without linking the personal shopping behavior to the person who bought that cash from the bank, to keep the feature of "anonymity of payer" of real money.

译文：银行必须具备能追踪电子现金是否被复制或重复使用的能力，而又不能将个人购买行为与从银行购现金的人联系起来，以保持真钱币的"匿名"特色。

本句中的"to track…"作定语，修饰宾语"ability"，"without…"是条件状语，"to keep…"是目的状语。

[3] These new types of bank server software, which adopt up the modern information technologies, can provide convenient, real-time, secure and reliable bank services.

译文：这些新型银行服务软件采用了现代信息技术，能提供方便、实时、安全和可靠的银行服务。

本句中"which"引导的是非限定定语从句，修饰主语，"convenient, real-time, secure and reliable"修饰宾语"bank services"。

[4] Research and Intelligence System (RIS) continually scans and monitors the environment, observing and drawing conclusions about the events that affect integrated logistics operations.

译文：研究和情报系统（RIS）不断地监控环境，观察并总结影响整合物流操作的事件。

本句的谓语是"scans and monitors"，现在分词短语"observing and drawing conclusions about the events that affect integrated logistics operations"作宾语补足语。

[5] The inter-firm environment includes elements in the external environment that directly affect the firm and over which the firm does exercise some control, such as the channel of distribution.

译文：公司间环境包括一些直接影响公司，且公司有一定控制的外部环境要素，如分销渠道。

本句的"that directly affect..."作定语，"such as the channel of distribution"是"elements in the external environment"的同位语。

6.4.2　Exercises

1．Translate the following phrases into English or Chinese

（1）electronic check

（2）stock quotation

（3）digital cash software

（4）fund transfer

（5）logistics manager

（6）电子钱包

（7）数字签名

（8）物流信息系统

（9）订单处理系统

（10）库存状态

2．Identify the following to be True or False according to the text

（1）The Internet payment means conducting fund transfer, money receiving and disbursing electronically.

（2）Peak logistics efficiency and effectiveness demand a superior integrated logistics information system.

（3）An ILIS gathers information from all possible sources to assist the integrated logistics manager in making decisions.

（4）The final subsystem of ILIS is the reports and outputs system.

（5）The advanced encryption and authentication system can't keep the feature of "anonymity of payer" of paper money.

（6）On Internet, the cost for serving one person is not the same as that for serving 10 thousand persons.

3．Reading Comprehension

（1）There are many kinds of digital money, or electronic token, which mainly could be classed into two categories: _____ and after time payment mechanism.

　　A．digital payment mechanism

　　B．check payment mechanism

　　C．real-time payment mechanism

　　D．credit card payment mechanism

(2) _____ is the set of activities necessary to make the correct goods ready for shipment to the customer—right up to the point where warehousing assembles the order.
A. Order processing
B. Reports processing
C. Outputs processing
D. Information processing

(3) Some electronic-check software can provide e-wallets for their customers, which also enable the customers to write checks at anywhere even they have no _____.
A. money
B. information
C. digital cash
D. bank account

(4) Since the digital cash is only a _____ so it needs to be verified as authentic.
A. number
B. digital number
C. data
D. file

(5) _____ are computer-based and provide solutions to complex integrated logistics problems using analytical modeling.
A. SSD
B. DSS
C. DDS
D. DSD

6.4.3 Reading Material

Electronic Marketing

By using the Internet, manufacturers can directly contact customers without using intermediaries. The manufacturer's direct marketing can be realized as long as they sell established brands and their home site is well known. If a manufacturer's site does not have high visibility, just opening a home page and passively waiting for customers' access may not contribute greatly to sales. Therefore, it is necessary for companies to heavily advertise their Web sites' address. Any cost-effective advertisement method can be employed for this purpose. One example is to link the site to well known electronic directories, and most manufacturers use the directory service of intermediaries. These intermediary sites are called electronic shopping malls.

Initially, the main concern for electronic marketing involved securing technologies necessary to implement Internet-based marketing, such as powerful search capability and secure electronic payment. However, today the main concern of management is shifting to how to utilize the opportunity of Internet-based marketing to enhance competitiveness in harmony with existing marketing channels.

Increasingly, companies are classifying customers into groups and creating targeted messages for each group. The sizes of these targeted groups can be smaller when companies are using the Web in some cases, just one customer at a time can be targeted. New research into the behavior of Web site visitors has even suggested ways in which Web sites can respond to visitors who arrive at a site with different needs at different times.

Most companies use the term "marketing mix" to describe the combination of elements that they use to achieve their goals for selling and promoting their products and services. When a company decides which elements it will use, it calls that particular marketing mix its marketing strategy. Companies—even those in the same industry—try to create unique presences in their markets. A company's marketing strategy is an important tool that works with its Web presence to get the company's message across to both its current and prospective customers.

Most marketing classes organize the essential issues of marketing into the four Ps of marketing: product, price, promotion, and place. Product is the physical item or service that a company is selling. The intrinsic characteristics of the product are important, but customers' perceptions of the product, called the product's brand, can be as important as the actual characteristics of the product.

The price element of the marketing mix is the amount the customer pays for the product. In recent years, marketing experts have argued that companies should think of price in a broader sense, that is, the total of all financial costs that the customer pays to obtain the product. This total cost is subtracted from the benefits that a customer derives from the product to yield an estimate of the customer value obtained in the transaction. The Web can create new opportunities for creative pricing and price negotiations through online auctions, reverse auctions, and group buying strategies. These Web-based opportunities are helping companies find new ways to create increased customer value.

Promotion includes any means of spreading the word about the product. On the Internet, new possibilities abound for communicating with existing and potential customers. Companies are using the Internet to engage in meaningful dialogs with their customers using e-mail and other means.

When a company takes its business to the Web, it can create a Web site that is flexible enough to meet the needs of many different users. Instead of thinking of their Web sites as collections of products, companies can build their sites to meet the specific needs of various types of customers. A good first step in building a customer-based marketing strategy is to identify groups of customers who share common characteristics. The use of customer-based marketing approaches was pioneered on B2B sites. B2B sellers were more aware of the need to customize product and service offerings to match their customers' needs than were the operators of B2C Web sites. In recent years, B2C sites have increasingly added customer-based marketing elements to their Web sites.

Advertising is an attempt to disseminate information in order to effect a buyer-seller transaction. The Internet redefined the meaning of advertising. The Internet has enabled consumers to interact directly with advertisers and advertisements. The Internet has provided the sponsors with two-way communication and e-mail capabilities, as well as allowing the sponsors to target specific groups on which they want to spend their advertising dollars, which is more accurate than traditional telemarketing. Finally, the Internet enables a truly one-to-one advertisement.

6.5 专业英语应用模块

6.5.1 Appointment Letter

1．预约信的内容与要求

在预约信中，你要请求约会并说明原因，需要给出建议的约会时间和地点等信息。如果你的时间比较充裕，预约时可给出你可接受的时间由对方决定。最后一般请对方答复并进行确认。在文字组织方面，只要做到清楚、简洁、礼貌就行。

预约信最重要的是两项内容：
- 在何时会面。
- 在何地会面。

这两者必须叙述的清楚无误。另外，可简单谈及预约具体做什么，预约一般都是开门见山的。

示例如下：

Sunday, February 21, 2016

Professor Adam,

　　I wonder if I could see you in your office next Monday morning, at 10:30. I have so many problems to talk over with you on my current paper on Environment Protection. Please leave a note for me on your door and let me know whether it is OK, or when a convenient time will be.

Wang Xiaolin

2．常见的预约用语

Could we meet at … on …?

Could you wait for me at … on …?

Shall we meet at …?

…, please meet me at …

I would like to see you …

I wonder if I could see you …

6.5.2 Letter of Inquiry

1．咨询信的内容与要求

咨询信是写信人就自己不太熟悉和不太了解或不理解的事情或问题，向有关部门或专家请求解答时所使用的一种专用书信。咨询信是请求解答疑问的书信，写信人由于在工作或学习上遇到难以解开的问题或不理解的事情时才会写咨询信。

咨询信的主要内容包括以下几个方面：
- 写信的目的。
- 介绍背景。
- 询问详情。
- 表示感谢，期待回复，询问进一步的信息。

示例如下：

Dir Sir/Madam,

  From your advertisement in the China Daily, March 15, I have known that your firm needs a programmer in your application development department. I would be interested in exploring the possibility of obtaining such a position within your firm.

  I received Bachelor of Science degree in computer science from Shanghai Industrial University in July 2014. Since then I have been working in the Great IT company as a programmer for C/C++ programmer whom you need. And I also have some experience in database administration. Details of my educational background and working experience are contained in the enclosed resume.

  May I schedule an appointment for an interview with you to discuss my qualifications in detail?

  I am looking forward to your reply.

  Your kind help would be greatly appreciated.

Yours sincerely.
Wang Xiang

2. 常用的咨询用语

I am writing to ask if …

I am writing to explore the possibility of obtaining the position in your firm.

I am looking forward to hearing from you.

I am looking forward to your prompt response.

Would it be possible to …

Would you mind doing something?

I would be grateful if you would …

I would appreciate it very much if you could consider my suggestions.

I should be most obliged to you if you could…

I wonder if you could …

Please send me all the relevant information concerning the courses you offer.

Thank you for your time and attention.

Computer Applications

7.1 Office Automation

7.1.1 Text

Office Automation (OA) is the application of computer and communications technology to improve the productivity of clerical and managerial office workers. In the mid-1950s, the term was used as a synonym for almost any form of data processing, referring to the ways in which bookkeeping tasks were automated. After some years of disuse, the term was revived in the mid-1970s to describe the interactive use of word and text processing systems, which would later be combined with powerful computer tools, thereby leading to a so-called "integrated electronic office of the future".[1]

Personal computer-based office automation software has become an indispensable part of electronic management in many countries. Word processing programs have replaced typewriters; spreadsheet programs have replaced ledger books; database programs have replaced paper-based electoral rolls, inventories and staff lists; personal organizer programs have replaced paper diaries; and so on.

Office automation encompasses six major technologies:
- Data processing—Information in numeric form usually calculated by a computer.
- Word processing—Information in text form—word and numbers.
- Graphics—Information that may be in the form of numbers and words, then keyed into a computer and displayed on a screen in a graph, chart, table, or other visual form that makes it easier to understand.
- Image—Information in the form of pictures. Here an actual picture or photograph is taken, entered into the computer, and shown on a screen.
- Voice—The processing of information in the form of spoken words.

● Networking—The linking together electronically of computer and other office equipment for processing data, words, graphics, image, and voice.

Initially, systems sold by major manufacturers were aimed at clerical and secretarial personnel. These were mainly developed to do word processing and record processing (maintenance of small sequential files, such as names and addresses, which are ultimately sorted and merged into letters). More recently, attention has also been focused on systems (Fig. 7-1), which directly support the principals (managers and professional workers). Such systems emphasize the managerial communications function.

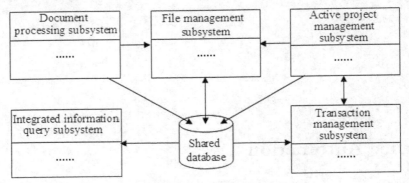

Fig. 7-1　Office Automation System

Today's organizations have a wide variety of office automation hardware and software components at their disposal. The list includes telephone and computer systems, electronic mail, word processing, desktop publishing, database management systems, two-way cable TV, office-to-office satellite broadcasting, on-line database services, and voice recognition and synthesis. Each of these components is intended to automate a task or function that is presently performed manually. But experts agree that the key to attaining office automation lies in integration—incorporating all the components into a whole system such that information can be processed and communicated with maximum technical assistance and minimum human intervention.

The fast pace of modern business requires critical information quickly. At the same time, government demands and business bureaucracy require extensive amounts of paperwork. As a result, modern business offices are reexamining traditional methods of doing office work to find better ways to capture and communicate information when and where it is needed.[2] They seek the most efficient method to generate, record, process, file, and communicate or distribute information. Modern technology offers office automation as the foundation of an economical solution. Present and future office system aim to develop integrated information processing networks that bring together everything a firm needs to conduct its daily business effectively.

（1）Word Processing

Word processing refers to the methods and procedures involved in using a computer to create, edit, and print documents. Most standard word processing features are supported, including footnotes and mail-merge but no tables or columns. The interface uses customizable toolbars, and the editing screen is a zoom-able draft mode that optionally displays headers, footnotes, and footers.

An un-editable print preview displays a full-page or facing-page view. Fonts, keyboard layouts, and input direction change when you select a new language, and keyboard layouts can be customized.[3]

Word processors vary considerably, but all word processors support some basic features. Word processors that support only the basic features (and maybe a few others) are called text editors. Most word processors, however, support additional features that enable you to manipulate and format documents in more sophisticated ways. These more advanced word processors are sometimes called full-featured word processors. Full-featured word processors usually support the following features:

- Insert and Delete text
- Cut, paste and copy
- Page size and margins
- Search and replace
- Word wrap
- Print
- File management
- Font specifications
- Footnotes and cross-references
- Graphics
- Headers, footers and page numbering
- Layout
- Macros
- Merge
- Spell checker
- Tables of contents and indexes
- Thesaurus
- Windows
- WYSIWYG (What You See Is What You Get)

（2）Electronic Spreadsheet

Microsoft Excel is a spreadsheet program that allows you to organize data, complete calculations, make decisions, graph data, develop professional looking reports, publish organized data on the Web, and access real-time data from Web sites.[4] Using Microsoft Excel, you can create hyperlinks within a worksheet to access other Office documents on the network, an organization's intranet, or the Internet.

Electronic spreadsheet can be used in all walks of life, for example, accountants use electronic spreadsheets to check financial statements and compile payrolls; in commerce, electronic spreadsheets are used to prepare budget and perform comparison on quoted prices; teachers use electronic spreadsheets to record marks of courses students get in examinations; scientists use electronic spreadsheets to analyze experimental data. Housewives use electronic spreadsheets to keep track of family expenditure.

In Excel, formulas and functions can be used to perform statistics or computations of data in electronic spreadsheets. When data change, the results of computations will automatically upgrade. All formulas begin with the sign =. It contains arithmetic operation signs, text operation signs, comparative operation sign and quoting operation sign—totally four categories of signs. Functions are built-in formulas pre-defined in Excel, including SUM, AVERAGE, COUNT, and MAX, etc.

Various statistic charts in electronic spreadsheets represent the data in unit squares, so as to make the data easy and intuitive to understand, meanwhile, when the data change, the charts automatically follow the change.

In Excel, charts include "imbedded chart" and "independent chart". The former is placed and displayed and printed together with worksheets; the latter is an independently generated worksheet, and it is printed separated from the original datasheet. In Excel, there are 2-dimensional chart and 3-dimentational charts. They are gradually generated by using the "chart direction" button in tool bar or "picture" command in "insert" menu. The user can edit the generated charts. There are zigzag chart, bar chart, pie chart and others available for selection.

Excel has not only abilities to perform simple data management and computation but also to set up database. To use datasheet, increase or delete record, and do sequencing, sieving, classifying and summing up data.

（3）Video Conferencing

Video conferencing involves the linking of remote sites by one-way or two-way television. If conference rooms in offices can be equipped with the necessary audiovisual facilities, travel time and money can be saved by holding a teleconference instead of a face-to-face conference.[5] Many businesses are experimenting with sales and board meetings through video conferencing. The cost is still high, especially if the video conference involves a direct two-way satellite channel connection.

Key Words

accountant	会计，账房，出纳
bureaucracy	官僚机构，官僚
disposal	支配权，使用权，配置
encompass	围绕，包含
expenditure	开销，支出
footnote	脚注
indispensable	不可缺少的，必需的
intervention	干预，介入
ledger	分类账，底账
managerial	管理人的，经理的
payroll	发薪簿，工资表
quote	报价，估价
revive	再流行，复苏
statistic	统计上的，统计学的

Notes

[1] After some years of disuse, the term was revived in the mid-1970s to describe the interactive use of word and text processing systems, which would later be combined with powerful

computer tools, thereby leading to a so-called "integrated electronic office of the future".

译文：经若干年搁置后，20 世纪 70 年代中期该词再次被用来描述字和文本处理系统的交互使用，这种系统后来又与强有力的计算机结合导致所谓的"未来综合电子办公室"的出现。

本句的"After some years of disuse"是时间状语，主句用被动语态，"which"引导的是非限定定语，"thereby leading…"是结果状语。

[2] As a result, modern business offices are reexamining traditional methods of doing office work to find better ways to capture and communicate information when and where it is needed.

译文：因此，现代商业办公室都在对传统的办公方法进行研究与调整，以找到可以随时随地获取信息和传递信息的较好的途径。

本句的"to find…"作目的状语。

[3] Fonts, keyboard layouts, and input direction change when you select a new language, and keyboard layouts can be customized.

译文：当你选择一种新语言时，字体、键盘布局和输入方向均可变化，并且键盘布局可以定制。

这是一个由"and"连接起来的并列句，本句的"when"引导的是时间状语。

[4] Microsoft Excel is a spreadsheet program that allows you to organize data, complete calculations, make decisions, graph data, develop professional looking reports, publish organized data on the Web, and access real-time data from Web sites.

译文：Excel 是一个电子表格程序，用于组织数据、完成计算、做出决策、将数据图表化、生成专业水准的报告、在 Web 上发布组织好的数据以及在 Web 站点上存取实时数据。

本句的主句是"Microsoft Excel is a spreadsheet program"，而"that"后的部分都作定语，修饰"a spreadsheet program"。

[5] If conference rooms in offices can be equipped with the necessary audiovisual facilities, travel time and money can be saved by holding a teleconference instead of a face-to-face conference.

译文：如果会议室内能安装所需的声像设备，通过举行远距离会议可代替面对面的会议，可节约旅行所花费的时间和经费。

本句由"If"引导条件状语从句，"by holding…"是方式状语。

7.1.2　Exercises

1. Translate the following phrases into English or Chinese

（1）voice recognition and synthesis

（2）statistic chart

（3）satellite broadcasting

（4）desktop publishing

（5）integrated information processing network

（6）键盘布局

（7）数据处理

（8）办公自动化

（9）字处理程序

（10）电子表格程序

2. Identify the following to be True or False according to the text

（1）In Excel, there are 2-dimensional chart and 4-dimentational charts.

（2）In Excel, charts include "imbedded chart" only.

（3）Modern technology offers OA as the foundation of an economical solution.

（4）The advanced word processors are sometimes called full-featured word processors.

（5）All formulas begin with the sign =.

（6）Word processing refers to the methods involved in using a typewriter to create and edit documents.

3. Reading Comprehension

（1）Personal computer-based office _____ has become an indispensable part of electronic management in many countries.

  A. automation hardware

  B. hardware

  C. software

  D. automation software

（2）Video conferencing involves the linking of remote _____ by one-way or two-way television.

  A. cities

  B. pages

  C. sites

  D. software

（3）_____ are built-in formulas pre-defined in Excel, including SUM, AVERAGE, COUNT, and MAX, etc.

  A. Functions

  B. Data

  C. Charts

  D. Tables

（4）In Excel, formulas and functions can be used to perform statistics or computations of data in _____.

  A. electronic spreadsheets

  B. sheets

  C. files

  D. documents

（5）Today's _____ have a wide variety of office automation hardware and software components at their disposal.

  A. family

  B. schools

  C. organizations

D. persons

7.1.3 Reading Material

3D Printing

You have heard of 3D printing from newscasters and journalists, astonished at what they have witnessed. A machine reminiscent of the star trek replicator, something magical that can create objects out of thin air. It can "print" in plastic, metal, nylon, and over a hundred other materials. It can be used for making nonsensical little models like the over-printed Yoda, yet it can also print manufacturing prototypes, end user products, quasi-legal guns, aircraft engine parts and even human organs using a person's own cells.

We live in an age that is witness to what many are calling the Third Industrial Revolution. 3D printing, more professionally called additive manufacturing, moves us away from the Henry Ford era mass production line, and will bring us to a new reality of customizable, one-off production (Fig. 7-2).

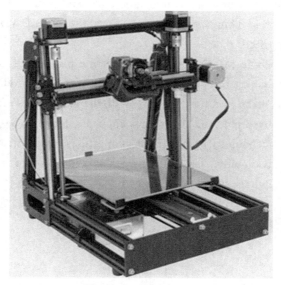

Fig. 7-2 3D printing

Need a part for your washing machine? As it is now, you'd order from your repairman who gets it from a distributer, who got it shipped from China, where they mass-produced thousands of them at once, probably injection-molded from a very expensive mold. In the future, the beginning of which is already here now, you will simply 3D print the part right in your home, from a CAD file you downloaded. If you don't have the right printer, just print it at your local foundry.

3D printers use a variety of very different types of additive manufacturing technologies, but they all share one core thing in common: they create a three dimensional object by building it layer by successive layer, until the entire object is complete. It's much like printing in two dimensions on a sheet of paper, but with an added third dimension: UP, the Z-axis.

Each of these printed layers is a thinly-sliced, horizontal cross-section of the eventual object. Imagine a multi-layer cake, with the baker laying down each layer one at a time until the entire cake is formed. 3D printing is somewhat similar, but just a bit more precise than 3D baking.

While most people have yet to even hear the term 3D printing, the process has been in use for decades. Manufacturers have long used the printers in the design process to create prototypes for traditional manufacturing. But until the last few years, the equipment has been expensive and slow.

Now, fast 3D printers can be had for tens of thousands of dollars, and end up saving the companies many times that amount in the prototyping process. For example, Nike uses 3D printers to create multi-colored prototypes of shoes. They used to spend thousands of dollars on a prototype and wait weeks for it. Now, the cost is only in the hundreds of dollars, and changes can be made instantly on the computer and the prototype reprinted on the same day.

Some companies are using 3D printers for short run or custom manufacturing, where the printed objects are not prototypes, but the actual end user product. As the speeds of 3D printing go up and the prices come down, look for more and more of this. And expect more availability of personally customized products.

Even if you don't design your own 3D model, you can still print some very cool pieces. There are model repositories such as Thingiverse, 3D Parts Database, and 3D Warehouse that have model files you can download for free.

What do all these people print? It is limitless. Some print things like jewelry, some print replacement parts for appliances such as their dishwasher, some invent all sorts of original things, some create art, and some make toys for their kids. With the many types of metal, plastic, glass, and other materials available (even gold and silver), just about anything can be printed.

This is a disruptive technology of mammoth proportions, with effects on energy use, waste, customization, product availability, art, medicine, construction, the sciences, and of course manufacturing. It will change the world as we know it.

7.2 Distance Education

7.2.1 Text

Computer-based approaches to education can enhance almost all other forms of distance education.[1] The ability of computer technology to interface with and control other technologies has placed it in the forefront of all technologies as the greatest proponent of change in the educational environment. As rapid advances in computer technology foster obsolescence of older equipment, the economic realities are that prices continue to decrease for the technology. It must be understood that the reason computer programs are praised as being innovative and productive in the educational arena, is often traced to the amount of time expertise utilized by the instructional designer in production of program. Computers have the same drawbacks as other media, and yet they also offer opportunities for counteracting the inherent deficits. The computer is just a tool, abate a very powerful one, that is available for educations in their pursuit to create an educational environment in which learning will take place.

The definition of computer-mediated communication appears to parallel the definition of distance education in that it often removes the teacher from the student in time and location.

Computer technologies have caused tremendous advancements in information storage and retrieval sciences that are combined with electronic communications to produce tremendous educational tools: Internet, telecommunication, electronic bulletin boards, E-mail, video conferences, and many others.

The development of videodisc and laser technology provided several unique features: ability to store large amounts of data, ability to display still images indefinitely without wear to the disc, and the ability to access any frame within microseconds. The almost instantaneous frame access and massive still storage ability make the videodisc a uniquely ideal visual storage format that can be used by educators and can easily be controlled by computers, providing an interactive media useful in all forms of education.

1. Internet and Distance Education

Recent developments and decreasing costs are providing access to unsurpassed amounts of information from around the globe through electronic Internet. All that is required for students anywhere in the world to access this information is a computer, modem, telephone, and access port (commercial or educational). The ability for students to conduct research without leaving their homes is changing the way that educational institutions structure research. Electronic publishing is one of the fastest growing fields, but establishment of standards before it will be accepted equally as printed material.

Telecommunications support distance education by providing delivery systems that carry programs and allow interaction between the participants. The availability of satellite and cable delivered programming provide economic options to school districts throughout the country. Programs such as the Jason Project and Space Explorer, allow any school to procure access to distance learning projects that can be seen world wide and interactive with each remote site. Rapid advancements in optical technology are providing methods of transferring increasing amounts of information and will allow remote students to use cable or telephone connections to use interactive multimedia programming.

Electronic bulletin boards and news services provide access to world wide discussions on any range of topics. These one-to-many communication platforms allow postings of assignments and course information for distance education. Group discussions allow individuals to analyze the thoughts of their peers as well as those of uncountable experts on any topic.

Increasing access to the Internet is promoting E-mail as a one-to-one or one-to-many platform of communication. E-mail allows peer-to-peer conversations and can be almost instantaneous in response if the user sets up a two-way conversation mode. The ability to attach text and graphic files to E-mail allows the user to send papers and articles to any number of addresses. E-mail is extremely useful for correspondence between teacher and student, allowing feedback to any questions that the student might encounter while working on course material and allowing the teacher to transmit grades and feedback on submitted lessons.

Often, the need for face-to-face interaction is required and can easily be allowed through the use of video teleconferencing. Students enter into a cooperative learning process, enhancing the instructional experience and thus reducing isolation. Two-way full-motion video is available through satellite communications, coaxial cable, and in the near future through fiber-optic link. The current cost of video conferencing restricts its use to mostly commercial purposes, but as the price continues to drop it will undoubtedly be used throughout most distance education courses.

All of the technologies discussed above are easily controlled through computer-mediated communications and many are extensions of computer technology. There are many benefits derived from the use of computer-based communications. The first is the ability of computer programs to be interactive and provide feedback to the student. The second value is the ability of computers to become any and all existing media, including books and musical instruments. A third is that information can be presented from many different perspectives. Further values include the ability to use computers in simulation models and the ability to engineer computers to be reflective.

2. Multimedia and Distance Education

The interactive multimedia materials can be delivered to the student via a number of methods. Some of these methods range from demonstrations and presentations, drills and practices, and tutorials, which are more teacher-centered methods on teaching and learning, to cooperative and collaborative learning, problem-solving and Web-based learning, which are more student-centered methods of teaching and learning.

In the teacher-centered learning methods, the teacher is the one in control of the information that is received by the students and is responsible for how much information is being disseminated to them. In the student-centered learning methods, the students construct their own knowledge and bring their authentic experiences into the learning process with the teacher as the facilitator.

With the teacher-centered approach, the students are given the same amount of information at the same time and able to use several senses (e.g., sight, sound and even interactive experiences), as in the case with presentations and demonstrations to process the information. They are also able to retain and recall the information as well as obtain mastery in the subject matter with drills, practices and tutorials, which are highly interactive.

With student-centered approach, the multimedia material can be used to foster team processing and active learning as with collaborative and cooperative methods, encourage higher-level learning skills and increase comprehension and retention rates, as in the case of using problem-solving methodologies. In the student-centric instructional model, the learners must play an active part in their learning and must construct their own knowledge or meaning of what they learn. The learners determine how to reach the desired learning outcomes themselves. This learning mode is embedded in the constructivist learning philosophy, which has evolved over the last half of the 20th C and has its foundations in cognitive learning psychology. In this constructivist model, learners must construct their own meaning of what they learn. The teacher, in this case, is no longer perceived as the sole authority of learning, but rather, as the person to facilitate learning, guiding and supporting learners' own construction of knowledge.[2]

In the present days, many are turning to multimedia as a means to better communicate instructional message and to foster better feedback on the information exchanged.[3] For many years, multimedia and multimedia developers were housed in selected industries such as advertising, entertainment, games and corporate Computer-Based Training (CBT) systems. However, multimedia is now penetrating the education field and changing the way teachers teach and students learn. With the advent of multimedia technology in the classrooms, teachers can equip themselves with these technological skills and become better communicators of their content materials, and thus enabling the students to learn in a more productive way. Thus, the use of the PC and digital multimedia technologies has given rise to new modes of learning and enabled new and innovative ways.

Furthermore, with the advent of the Internet and the World Wide Web, in the mid-1990s, a revolutionary transformation is taking place in our educational methodologies.[4] This will extensively widen our scope of learning into a global perspective and connect our learners to educational resources and information worldwide. Web-based learning and distance learning over the Internet are now available to those who want them.[5] It is now possible for anyone with an Internet connection to access the innumerable libraries and information resources of the world, changing the landscape in the education field into an IT-oriented one.

The arrival of the digital technologies has been a boon to the educational field, and has led, in recent years, to the institutions of higher learning in China rapidly embracing digital multimedia technology in their educational curricula. The use of modern technology on beefing up the delivery of learning materials in our education system must reflect the changing times. Consequently the educators at higher institutions of learning are facing a new challenge today, i.e., to integrate these multimedia technologies into the classroom to enhance the teaching and learning environments for both the teacher and the students.

Key Words

abate	减轻，废除
arena	舞台，场地
assignment	分配，转让
attach	依附上，使喜爱
challenge	挑战
coaxial	共轴的
cognitive	认知的，认识的
counteract	抵消，阻碍
deficit	赤字
encounter	相遇，遭遇
enhance	提高，加强
forefront	最前部，最前线
foster	培养，鼓励
innovative	改革，创新
instantaneous	即时的，瞬时
isolate	使孤立，使隔离
massive	宽大的，宏伟的
mastery	精通，掌握
obsolescence	荒废，淘汰
outcome	结果，出口

philosophy	哲学，哲理
procure	获得，取得
psychology	心理学
pursuit	追踪
submit	委托，提交
telecommunication	电信，远距离通信
tutorial	学习手册，学习指导
uncountable	无数的，不可数的
uniquely	独特地，稀罕地
unsurpassed	非常卓越的

Notes

[1] Computer-based approaches to education can enhance almost all other forms of distance education.

译文：将计算机技术用于教育改善了几乎所有其他远程教育的方式。

本句的"to education"是主语补足语，说明主语，"distance education"是定语，修饰宾语。

[2] The teacher, in this case, is no longer perceived as the sole authority of learning, but rather, as the person to facilitate learning, guiding and supporting learners' own construction of knowledge.

译文：在这种情况下，老师不再被认为是唯一权威的教学者，而更可能是一个促进、指导及支持学生自有知识体系的人。

本句的"in this case"是插入语，"but rather"也是插入语，表示转折，"as the sole…"和"as the person…"是方式状语。

[3] In the present days, many are turning to multimedia as a means to better communicate instructional message and to foster better feedback on the information exchanged.

译文：如今，多媒体技术作为一种能很好地传达指导意见和促进对所交换信息更好的反馈的方法，很多领域的目光都转向了它。

本句的"In the present days"是时间状语，"as a means…"作宾语补足语。

[4] Furthermore, with the advent of the Internet and the World Wide Web, in the mid-1990s, a revolutionary transformation is taking place in our educational methodologies.

译文：此外，随着20世纪90年代中期互联网和万维网的发展，在教育方法论领域产生了革命性的转变。

本句的"with the advent of the Internet and the World Wide Web"是条件状语，"in the mid-1990s"是时间状语，"in our educational methodologies"也是状语。

[5] Web-based learning and distance learning over the Internet are now available to those who want them.

译文：网络教学和基于互联网的远程教学使需要它们的人都能够享用。

本句的主语是"Web-based learning and distance learning"，"over the Internet"是主语补足语。

7.2.2 Exercises

1. Translate the following phrases into English or Chinese

（1）distance education

（2）information storage
（3）optical technology
（4）news service
（5）educational tools
（6）电子公告板
（7）电子通信
（8）仿真模型
（9）通信平台
（10）点对点

2. Identify the following to be True or False according to the text

（1）The interactive multimedia materials cannot be delivered to the student via a number of methods.

（2）The development of videodisc and laser technology could not provide help to distance education.

（3）The arrival of the digital technologies has been a boon to the educational field.

（4）The availability of satellite and cable delivered programming provide economic options to school districts throughout the country.

（5）The definition of computer-mediated communication appears to parallel the definition of distance education.

（6）There are many benefits derived from the use of computer-based communications.

3．Reading Comprehension

（1）The ability of _____ to interface with and control other technologies has placed it in the forefront of all technologies as the greatest proponent of change in the educational environment.

　　A. computer technology
　　B. electronic technology
　　C. technology
　　D. industry technology

（2）The ability for _____ to conduct research without leaving their homes is changing the way that educational institutions structure research.

　　A. students
　　B. programmers
　　C. users
　　D. manager

（3）Telecommunications support distance education by providing _____ that carry programs and allow interaction between the participants.

　　A. analysis systems
　　B. delivery systems
　　C. design systems
　　D. operate systems

(4) Increasing access to the Internet is promoting E-mail as a _____ platform of communication.

A. one-to-one or many-to-many
B. one-to-many
C. one-to-one or one-to-many
D. many-to-many

(5) With the teacher-centered approach, the _____ are given the same amount of information at the same time and able to use several senses.

A. user
B. scholar
C. programmer
D. students

7.2.3 Reading Material

Multimedia Technology

It is necessary for a multimedia system to support a variety of media types. This could be as modest as text and graphics or as rich as animation, audio and video. However, this alone is not sufficient for a multimedia environment. It is also important that the various sources of media types are integrated into a single system framework. A multimedia system is the one which allows end users to share, communicate and process a variety of forms of information in an integrated manner.

(1) Text

Many multimedia applications are based on the conversion of a book to a computerized form. This conversion gives the users immediate access to the text and lets them display pop-up windows, which give definitions of certain words. Multimedia applications also enable the user to instantly display information related to a certain topic that is being viewed. Most powerfully, the computerized form of a book allows the user to look up information quickly, without referring to the index or table of contents.

(2) Audio

The integration of audio sound into a multimedia application can provide the user with information not possible through any other method of communication. Some types of information can't be conveyed effectively without using sound. It is nearly impossible, for example, to provide an accurate textual description of the beat of a heart or the sound of the ocean. Audio sound can also reinforce the user's understanding of information presented in another type of media.

Audio sound is available in several different formats. One of them is the Windows wave file. A wave file contains the actual digital data used to play back the sound as well as a header that provides additional information about the resolution and playback rate. Wave files can store any type of sound that can be recorded by a microphone. Another type of audio sound that may be used is known as the Musical Instrument Digital Interface, or MIDI for short. The MIDI format is actually a specification invented by musical instrument manufactures. Rather than being a digitized form of the sound, the MIDI specification is actually a set of messages that describes what musical note is being played. The MIDI specification can't store anything except in the form of musical notes.

(3) Graphics Images

Static graphics images are an important part of multimedia because humans are visually oriented. Computational image processing can be defined as the operation of mathematical functions on numeric representations of pictorial scenes. In general it is part of an overall process of visual perception, pattern recognition and image understanding. These form the essential components of computer vision. The term digital image processing generally refers to processing of a two-dimensional picture by a digital computer. In a broader context, it implies digital processing of any two-dimensional data. A digital image is an array of real or complex numbers represented by a finite number of bits. An image given in the form of a transparency, slide, photograph, or chart is first digitized and stored as a matrix of binary digits in computer memory. This digitized image can then be processed and displayed on a high resolution television monitor.

(4) Animation and Full-motion Video

Animation refers to moving graphics images. Just as a static graphics image is a powerful form of communication, such is the case with animation. Animation is especially useful for illustrating concepts that involve movement. Full-motion video, such as the images portrayed in a television, can add even more to a multimedia application.

7.3 Artificial Intelligence

7.3.1 Text

Definitions of artificial intelligence vary along two main dimensions. One is concerned with thought processes and reasoning, the other addresses behavior. Also, some definitions measure success in terms of human performance, whereas the others measure against an ideal concept of intelligence, which we will call rationality. A system is rational if it does the right thing. All this gives us four possible goals to pursue in artificial intelligence: systems that think like humans, systems that think rationally, systems that act like humans and systems that act rationally (Fig. 7-3).

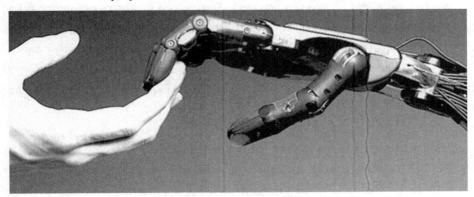

Fig. 7-3　Artificial intelligence

1. the Goals of Artificial Intelligence

(1) Acting humanly

The Turing Test, proposed by Alan Turing, was designed to provide a satisfactory operational definition of intelligence. Turing defined intelligent behavior as the ability to achieve human-level performance in all cognitive tasks, sufficient to fool an interrogator. Roughly speaking, the test he proposed is that the computer should be interrogated by a human via a teletype, and passes the test if the interrogator cannot tell if there is a computer or a human at the other end.[1] Programming a computer to pass the test provides plenty to work on. The computer would need to possess the following capabilities: natural language processing to enable it to communicate successfully in English (or some other human language), knowledge representation to store information provided before or during the interrogation, automated reasoning to use the stored information to answer questions and to draw new conclusions and machine learning to adapt to new circumstances and to detect and extrapolate patterns.

(2) Thinking humanly

If we are going to say that a given program thinks like a human, we must have some way of determining how humans think.[2] We need to get inside the actual working of human minds. There are two ways to do this: through introspection—trying to catch our own thoughts as they go by—or through psychological experiments. Once we have a sufficiently precise theory of the mind, it becomes possible to express the theory as a computer program.[3] If the program's input/output and timing behavior matches human behavior, that is evidence that some of the program's mechanisms may also be operating in humans. The interdisciplinary field of cognitive science brings together computer models from artificial intelligence and experimental techniques from psychology to try to construct precise and testable theories of the working of the human mind.

(3) Thinking rationally

The development of formal logic in the late 19th C and early 20th C provided a precise notation for statements about all kinds of things in the world and the relations between them.[4] By 1965, programs existed that could, given enough time and memory, take a description of a problem in logical notation and find the solution to the problem (if one exists).

There are two main obstacles to this approach. First, it is not easy to take informal knowledge and state it in the formal terms required by logical notation, particularly when the knowledge is less than 100% certain. Second, there is a big difference between being able to solve a problem "in principle" and doing so in practice. Even problems with just a few dozen facts can exhaust the computational resources of any computer unless it has some guidance as to which reasoning steps to try first.

(4) Acting rationally

Acting rationally means acting so as to achieve one's goals, given one's beliefs. An agent is just something that perceives and acts. In this approach, artificial intelligence is viewed as the study and construction of rational agents. In the "laws of thought" approach to artificial intelligence, the whole emphasis was on correct inferences. Making correct inferences is sometimes part of being a rational agent, because one way to act rationally is to reason logically to the conclusion that a given action will achieve one's goals, and then to act on that conclusion. On the other hand, correct inference is not all of rationality, because there are often situations where there is no provably correct thing to do, yet something must still be done.

2. Applications of Artificial Intelligence

(1) Speech recognition

In the 1990s, computer speech recognition reached a practical level for limited purposes. Thus United Airlines has replaced its keyboard for flight information by a system using speech recognition of flight numbers and city names. It is quite convenient. On the other hand, while it is possible to instruct some computers using speech, most users have gone back to the keyboard and the mouse as still more convenient.

(2) Understanding natural language

Just getting a sequence of words into a computer is not enough. Parsing sentences is not enough either. The computer has to be provided with an understanding of the domain the text is about, and this is presently possible only for very limited domains.

(3) Game playing

You can buy machines that can play master level chess for a few hundred dollars. There is some artificial intelligence in them, but they play well against people mainly through brute force computation—looking at hundreds of thousands of positions. To beat a world champion requires brute force being able to look at 200 million positions per second.

(4) Expert systems

A "knowledge engineer" interviews experts in a certain domain and tries to embody their knowledge in a computer program for carrying out some task. How well this works depends on whether the intellectual mechanisms required for the task are within the present state of artificial intelligence. When this turned out not to be so, there were many disappointing results. One of the first expert systems was MYCIN in 1974, which diagnosed bacterial infections of the blood and suggested treatments. It did better than medial students or practicing doctors, provided its limitations were observed. Namely, its ontology included bacteria, symptoms, and treatments and did not include patients, doctors, hospitals death, recovery, and events occurring in time. Its interactions depended on a single patient being considered. Since the experts consulted by the knowledge engineers knew about patients, doctors, death, recovery, etc., it is clear that the knowledge engineers forced what the experts told them into a predetermined framework. In the present state of artificial intelligence, this has to be true. The usefulness of current expert systems depends on their users having common sense.

3. Robotics

Most of artificial intelligence will eventually lead to robotics. Most neural networking, natural language processing, image recognition, speech recognition/synthesis research aims at eventually incorporating their technology into the epitome of robotics—the creation of a fully humanoid robot. The field of robotics has been around nearly as long as artificial intelligence.

According to the Oxford Dictionary, a robot is an "apparently human automaton, intelligent and obedient but impersonal machine".[5] Indeed, the word robot comes from robota, Czech for "forced labor". Yet, as robotics advances this definition is rapidly becoming old. Basically, a robot is a machine designed to do a human job (excluding research robots) that is either tedious, slow or hazardous. It is only relatively recently that robots have started to employ a degree of artificial intelligence in their work—many robots required human operators, or precise guidance throughout their missions. Slowly, robots are becoming more and more autonomous. The difference between robots and machinery is the presence of autonomy, flexibility and precision. Indeed, many robots are mere extensions of machinery—but as the field advanced more and more, the difference will be clearer.

Robotics is slowly making its way into the home. Recently, robotics released the world's first true personal robot—Cye. Cye allows its human operator to create a map of the environment (using a Windows interface) and download it via an Industrial Robot link to the robot. The robot will then be able to navigate the area doing various tasks—including vacuuming.

Robotics is an absolutely fascinating field that interests most people. As research from more serious robotics projects filter down into the commercial arena we should look forward to some very interesting (and cheap) virtual pets. Hopefully, commercial home-based robots will also be available for a price not more than an expensive vacuum cleaner. With computers becoming more and more powerful, interfacing home robots with your computer will become a reality, and housework will disappear.

Key Words

apparently	清楚地，显然地
bacterial	细菌的
behavior	行为，举止
conclusion	结论
extrapolate	推断，外推
guidance	指导，领导
hazardous	危险地，冒险的
interrogator	询问者
introspection	内省，反省
obedient	服从的，顺从的
obstacle	妨碍，障碍
ontology	存在论，本体论
pattern	模范，典型
rationality	合理性
recognition	识别
synthesis	合成
vacuum	真空
whereas	然而，鉴于

Notes

[1] Roughly speaking, the test he proposed is that the computer should be interrogated by a human via a teletype, and passes the test if the interrogator cannot tell if there is a computer or a

human at the other end.

译文：粗略地说，他提出的试验是计算机要由一个人通过电传打字机来向其提出问题，如果这个人不知道另一端是人还是计算机的话，那么该试验就算通过了。

本句的"Roughly speaking"是插入语，"that the computer…"是表语从句，"if"引导的是条件状语。

[2] If we are going to say that a given program thinks like a human, we must have some way of determining how humans think.

译文：如果我们说一个给定的程序能像人一样思维的话，我们必须有一些方法能确定人是怎样思维的。

本句由"if"引导条件状语，"determining how humans think"作定语，修饰"way"。

[3] Once we have a sufficiently precise theory of the mind, it becomes possible to express the theory as a computer program.

译文：一旦我们有了一个足够准确的智能理论，就有可能把该理论表示为计算机程序。

本句由"Once"引导时间状语从句，"it"指代后面的不定式短语。

[4] The development of formal logic in the late 19th C and early 20th C provided a precise notation for statements about all kinds of things in the world and the relations between them.

译文：19世纪后期和20世纪初期形式逻辑的发展为世界上所有事物及彼此之间关系的陈述提供了一套精确的符号。

本句的"formal logic"作定语，修饰主语"The development"；而"in the late 19th C and early 20th C"是时间状语；"for"后面的是宾语补足语。

[5] According to the Oxford Dictionary, a robot is an "apparently human automaton, intelligent and obedient but impersonal machine".

译文：按照牛津字典的解释，机器人是"外观像人的自动化装置，是智能的、服从的但不受感情影响的机器"。

本句的"According to the Oxford Dictionary"作状语，"is"后面的部分都是表语。

7.3.2　Exercises

1. Translate the following phrases into English or Chinese
　（1）human performance
　（2）thought process
　（3）image recognition
　（4）cognitive science
　（5）logical notation
　（6）神经网络
　（7）语音识别
　（8）人工智能
　（9）专家系统
　（10）虚拟宠物

2. Identify the following to be True or False according to the text
　（1）Thinking rationally means acting so as to achieve one's goals, given one's beliefs.

(2) Most of artificial intelligence will eventually lead to robotics.

(3) A system is rational if it does the right thing.

(4) Commercial home-based robots will also be available for a price not more than an expensive vacuum cleaner.

(5) You can play machines that can play master level chess for a few hundred dollars.

(6) Robotics is quickly making its way into the home.

3. Reading Comprehension

(1) In the "laws of thought" approach to artificial intelligence, the whole emphasis was on _____.

A. inferences
B. incorrect inferences
C. wrong inferences
D. correct inferences

(2) In the _____, computer speech recognition reached a practical level for limited purposes.

A. 1960s
B. 1980s
C. 1990s
D. 1970s

(3) One of the first _____ was MYCIN in 1974, which diagnosed bacterial infections of the blood and suggested treatments.

A. AI systems
B. expert systems
C. robots
D. home robots

(4) The difference between robots and machinery is the presence of _____.

A. autonomy, flexibility and precision
B. autonomy
C. flexibility
D. precision

(5) With computers becoming more and more powerful, interfacing home robots with your computer will become a reality, and _____ will disappear.

A. homework
B. task
C. work
D. housework

7.3.3 Reading Material

Intelligent Technologies

Artificial intelligent technologies attempt to understand intelligent entities because the constructed intelligent entities are interesting and useful in their own right. It is clear that computers with human-level intelligence would have a huge impact on our everyday lives and on the future course of civilization. The most common artificial intelligent technologies which we are going to analyse are: Natural Language, Robotics, Perceptive Systems, Expert Systems, Neural Networks and Intelligent Software.

1. Natural Language

Natural language is languages, including idioms that are used by humans. Natural language focuses on computer speech recognition and speech generation. The basic goal is to build computer hardware and software that can recognize human speech and "read" text and that can speak and write as well. A related goal is to build software that can perform research requested by humans.

2. Robotics

The goal of robotics research is to develop physical systems that can perform work normally done by humans, especially in hazardous or lethal environment.[1] Modern robotics is concerned with the development of numerically controlled machine tools and industrial fabrication machine that are driven by CAM (Computer-Aided Manufacturing) system.

3. Perceptive Systems

Like humans, robots need eyes and ears in order to orient their behavior. Since World War II, computer scientists and engineers have worked to develop perceptive systems, or sensing devices that can see and hear in the sense of recognizing patterns.[2] This filed of research, which is sometimes called "pattern recognition", has focused largely on military applications such as photo reconnaissance and missile control and navigation. Progress has been uneven because of problems teaching computers the differences between decoys and the real thing.

4. Expert Systems

Expert systems are relatively recent software applications that seek to capture expertise in limited domains of knowledge and experience and apply this expertise to solving problems. Media attention has perhaps focused more on expert systems than on any other member of the AI family. In part this is because such systems can assist the decision-making of managers and professionals when expertise is expensive or in short supply.

5. Neural Networks

People have always dreamed of building a computer that thinks, a "brain" modeled in some sense on the human brain. Neural networks are usually physical devices that electronically emulate the physiology of animal or human brains.

6. Intelligent Software

(1) Fuzzy Logic

Fuzzy Logic, a relatively new, rule–based development in AI, consists of a variety of concepts and techniques for representing and inferring knowledge that is imprecise, uncertain, or unreliable. Fuzzy logic can create rules that use approximate or subjective values and incomplete or ambiguous data. By allowing expressions such as "tall", "very tall", and "extremely tall", fuzzy logic enables the computer to emulate the way people actually make decisions, as opposed to defining problems and solutions using restrictive IF-THEN rules.

（2）Genetic Algorithms

Some artificial intelligence technologies are using problem-solving approaches found in nature. Genetic algorithms are one example. They consist of a variety of problem-solving techniques based on Darwinian principles of evolution. The algorithms start with building blocks that use processes such as reproduction, mutation, and natural selection to "breed" solutions. As solutions alter and combine, the worst ones are discarded, and better ones survive to go on and electronically breed with others to produce even better solutions. The process may produce results superior to anything crafted by humans.

（3）Intelligent Agents

The concept of intelligent agents evolved in 1950s as an offshoot of investigations into artificial intelligence. Since that time, the level of interest in intelligent software agents has grown. The tasks that an intelligent agent performs for a user require the characteristics of specificity, repetitiveness, and predictability. Agents can also work for a business process or a software application and easily outperform humans when they function at a high level.

7.4 专业英语应用模块

7.4.1 Notification Letter

如果你有什么重要的事情要告诉对方，可以通过 E-mail 等方式通知。通知范围很广，如工厂的迁址、开业、亲友的订婚、生日等，它可以是好消息，也可能是坏消息。不过在一些敏感的问题上要注意措辞。开会的通知要写清开会的时间、地点、参加会议的对象以及开什么会，还要写清要求。布置工作的通知，要写清所通知事件的目的、意义以及具体要求和做法等。

1．称呼

一般可以直接用收件人的姓名，也可以用 dear friends/clients/patients 等。

示例如下。

Dear Sir/Madam:

On March 11, we will be closing our branch office at 600 KaiXuan Road and moving to larger quarters in the Golden Building, 123 MeiLong Road, Suite 2120.

There is ample free parking at the new location and very convenient traffic tool: subway. If you have any question about us, please call, the office phone number will remain 021-62512678.

Your Truly
Computer Website Designer

2．常用的通知用语

I am happy to inform you that …

I am glad to tell you that …

Please be advised of my intention to resign from …

I wanted you to be the first to know ...

I have an announcement to make which is guaranteed to surprise you ...

It gives me great pleasure to announce ...

We have the pleasure of informing you that ...

7.4.2　Letters of Apologies

1．道歉信的内容与要求

　　道歉信是因为自己的原因或者工作失误等情况，可能引起对方不方便或者不愉快，以表示赔礼道歉，消除误会，增进友谊和信赖的信函。

　　在道歉信中，开头要简单交代对何事进行道歉，然后要向对方陈述事情发生的原因，事情原委要解释清楚，实事求是，简明扼要，以消除误会或矛盾。在最后部分表示遗憾和歉意，表明愿意补救的愿望，提出建议或者安排。在行文中注意态度要诚恳，用词要委婉，语气要温和、得体。

　　道歉信的主要内容体现在如下两点：
- 为什么而道歉；
- 有什么补救措施。

示例如下：

Dear Rose,

　　I am writing to say sorry that I can't meet you at the railway station on time.

　　I was glad to hear that you would come to see me and I thought that must go to the railway station to meet you. But yesterday night I was informed that I must attend an important business meeting on the day when you arrive. I will give a speech at the meeting which is on a project of which I am the leader. The meeting is supposed to be over at 11:00 am, which will be an hour later than your arrival time.

　　Being unfamiliar with the city, please wait for me at the railway station lounge with reading newspapers while waiting. I will go there as quickly as possible after the meeting.

　　I am sorry again for the inconvenience.

Best wishes,
Tom

2．常用道歉用语

I am terribly sorry I couldn't ...

I am sorry to say that ...

I must apologize about (not) doing something.

I am waiting to say sorry for ...

I feel so sorry that ...

I would like to express my apologies for ...

Once again, I am sorry for any inconvenience caused.

Please accept my apologies once more.

参考译文

第 2 章 硬件知识

2.1 中央处理器

课文

1. 什么是处理器?

处理器是解释并执行指令的功能部件。每个处理器都有一个独特的诸如 ADD、STORE 或 LOAD 这样的操作集,这个操作集就是该处理器的指令系统。计算机系统设计者习惯将计算机称为机器,所以该指令系统有时也称作机器指令系统,而书写它们的二进制语言叫作机器语言。注意,不要将处理器的指令系统与 BASIC 或 PASCAL 这样的高级程序设计语言中的指令相混淆。

指令由操作码和操作数组成,操作码指明要完成的操作功能,而操作数则表示操作的对象。例如,一条指令要完成两数相加的操作,它就必须知道:(1)这两个数是什么?(2)这两个数在哪儿?当这两个数存储在计算机内存中时,则应有指明其位置的地址,所以如果操作数表示的是计算机内存中的数据,则该操作数叫作地址。处理器的工作就是从存储器中取出指令和操作数,并执行每个操作,完成这些工作后就通知存储器送来下一条指令。处理器以惊人的速度一遍又一遍地重复以上这一步步的操作。

CPU 即中央处理器,是计算机的心脏(见图 2-1 略)。微机上的 CPU 实际上是一个很小的集成电路芯片。虽然大多数 CPU 芯片比一块眼镜片还小,但所包含的电子元件在几十年前却要装满一个房间。应用先进的微电子技术,制造者能够把上万个电子元件集成到很小很薄的硅芯片上,这些芯片的工作性能可靠且不费电。

中央处理器协调计算机各个部件的所有活动。它确定应该以什么顺序执行哪些操作。中央处理器也可取出存储器的信息并将操作结果存到存储媒体里,以备以后参考。

计算机的基本工作是处理信息。为此,计算机可以定义为接收信息的装置。信息是以指令和字符形式出现的,其指令组称为程序,字符则称为数据。该装置可对信息进行算术和逻辑运算,然后提供运算结果。程序的作用是指示计算机如何工作,而数据则是为解决问题提供的所需要的信息，两者都存储在存储器里。

人们认为计算机具有很多显著的功能。不过大多数计算机，无论是大型机还是小型机，都具有三个基本性能。

第一，计算机具有进行加、减、乘、除及求幂等各种算术运算的电路。

第二，计算机具有与用户通信的功能。如果我们不能输入信息和取出结果，这种计算机毕竟不会有多大用处。

第三，计算机具有进行判断的电路。计算机的电路能做出的判断为：一个数是否小于另一个数？两个数是否相等？一个数是否大于另一个数？

处理器由两个功能部件（控制部件和算术逻辑部件）和一组称作寄存器的特殊工作单元组成。

2．控制单元

控制单元是负责监督整个计算机系统操作的功能部件。在有些方面，它类似于智能电话交换机。因为它将计算机系统的各功能部件连接起来，并根据当前执行程序的需要控制每个部件完成操作。

可以说控制单元就是整个计算机的管理者，它协调着 CPU 的全部动作。控制单元用一个指令指针来跟踪将要处理的指令序列。控制单元在指针的引导下，按顺序从 RAM 中取出每条指令，并把它放在一个特定的指令寄存器中，然后控制单元将解释该指令以确定需要完成的操作。根据指令的解释结果，控制单元将信号发送到数据线以便从 RAM 中获取数据，然后再传送到 ALU 以完成一个处理过程。

3．算术逻辑单元

算术逻辑单元（ALU）是为计算机提供逻辑及计算能力的功能部件。控制部件将数据送到算术逻辑单元中，然后由算术逻辑单元完成执行指令所需要的算术或逻辑操作。

算术逻辑单元执行算术和逻辑操作。算术运算包括加、减、乘和除。逻辑操作包括比较，就是简单地要求计算机确定两个数是相等，还是一个数比另一个数大或者小。这些看起来似乎是很简单的操作。然而，通过这些操作的组合，算术逻辑单元就可以执行非常复杂的任务。例如，视频游戏就要使用算术运算和比较操作来确定显示在屏幕上的内容。

4．寄存器

寄存器是处理器内部的存储单元。控制部件中的寄存器用来跟踪正在运行程序的所有状态，它存储如当前指令，下一条将执行指令的地址以及当前指令的操作数这样一些信息。在算术逻辑单元中，寄存器存放要进行加、减、乘、除及比较运算的数据项。而其他寄存器则存放算术及逻辑运算的结果。

5．取指令周期和执行周期

在每个指令周期的开始，CPU 都从内存中取一条指令。典型的 CPU 中，用称作程序计数器的寄存器（PC）跟踪下一条被取出指令的地址。例如，考虑在计算机中每条指令占用 16 位字长的内存单元。假定程序计数器指向单元 300，那么 CPU 将到地址为 300 的单元取指令。在接下来的指令周期里，CPU 将依次从 301、302、303 等地址单元取指令。

被取出来的指令放入 CPU 的指令寄存器（IR）中，指令以二进制编码的形式存在，指明 CPU 将要采取什么操作。CPU 解释指令并完成要求的操作。大体上，这些操作被划分为以下 4 种类型。

(1) CPU—内存

数据可以从 CPU 传递到内存也可以由内存传递到 CPU。

(2) CPU—I/O

通过 CPU 与 I/O 模块之间的传递，数据可以输出或从外界输入。

(3) 数据处理

CPU 可以对数据执行一些算术或者逻辑操作。

(4) 控制

一条指令可以指定更改执行的顺序。例如，CPU 可能从 149 地址单元中取一条指令，该指令指明下一条指令从 182 地址单元读取。CPU 通过将程序计数器的值改变为 182 来记住这个事实。因此在下一个指令周期，指令将从 182 地址单元中取出而不是从 150 地址单元中取出。当然一条指令的执行可能会包括这些操作的组合。

6．CPU 的性能指标

(1) 时钟频率

计算机有一个发出脉冲以控制所有系统操作同步的系统时钟。系统时钟与保存每天时间的"实时时钟"不同。系统时钟设置数据传输和指令执行的速度或频率。系统时钟的频率决定了计算机执行指令的速度，因此限制了计算机在一定时间内所能执行的指令数。完成一个指令周期的时间用兆赫兹（MHz），或每秒百万个周期表示。

最初 IBM PC 机的微处理器时钟频率是 4.77MHz。现在的处理器的执行速度可以超过 2GHz。如果其他条件一样，时钟频率越高，就意味着处理速度越快。

(2) 字长

字长是指中央处理器（CPU）可以同时处理的位数。字长由 CPU 的寄存器大小和总线的数据线根数所决定。例如，字长为 8 位的 CPU 被称为 8 位处理器，它的寄存器是 8 位宽，可以同时处理 8 位数据。

字长较长的计算机在一个指令周期中要比字长短的计算机处理更多的数据。每个周期内处理的数据越多，处理器的性能就越高。例如，最初微型计算机的微处理器是 8 位，而现在都是 32 位或 64 位的微处理器。

(3) 高速缓冲存储器

高速缓冲存储器是影响 CPU 性能的另一个因素。高速缓冲存储器是一个特别的高速存储器，可以使 CPU 非常迅速地访问数据。高速 CPU 处理指令的速度非常快，以至于大部分时间不得不等待从处理速度很慢的 RAM 传送数据，这影响了处理速度。高速缓冲存储器可以保证一旦 CPU 请求就可以迅速获得数据。

阅读资料

计算机硬件基础

计算机是一种能接收、存储和处理数据，并能在存储指令程序控制下产生输出结果的快速、精确的符号加工系统。图 2-2 显示出了计算机系统的基本组成。系统的主要部件包括中央处理器、存储器、输入设备和输出设备。现在详细介绍每一个部件。

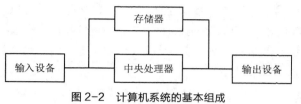

图 2-2　计算机系统的基本组成

1．中央处理器

处理器是计算机的"大脑"，它有能力执行用户提交给系统的指令或程序。处理器是实现加法和减法运算以及执行简单逻辑运算的地方。在大型计算机系统中，处理器被称作中央处理单元或 CPU，而在微机系统中，通常叫作微处理器。一台典型个人计算机的 CPU 由两部分组成：算术/逻辑运算单元和控制单元。不仅个人计算机如此，各种规模的计算机的 CPU 也都有这两部分。

2．存储器

存储器是计算机的工作区。存储器芯片有两种类型：只读存储器（ROM）和随机访问存储器（RAM）。只读存储器芯片是为数据（此数据可包括程序指令）只能读取的应用而设计的。这些芯片在加入系统之前，就已经由某个外部编程器装好数据了。这个工作一旦完成，其数据通常不再改变。即使在芯片断电以后，ROM 芯片也总能保存数据。随机访问存储器也称为读/写存储器，用来存储可以改变的数据。与 ROM 不同，RAM 芯片一旦掉电，数据就会丢失。许多计算机系统，包括个人计算机，都同时拥有 ROM 和 RAM。

3．输入设备

大部分输入设备以相似的方式工作。输入设备把接收到的信息和信号进行加工，使之变成 CPU 能处理的编码，然后送给 CPU。输入设备不但能给 CPU 输送信息，而且也能像开关控制电灯那样激活或终止处理过程。

4．输出设备

输出设备能告诉用户处理的结果和警告用户程序或操作在哪里错了。最普通的输出设备有显示器、点阵打印机、喷墨打印机、激光打印机、绘图仪和扬声器等。它们也以相似的方式工作，把 CPU 产生的编码形式的信号解码，变成人们易懂和易用的形式并显示出来。

5．系统总线

计算机的部件就是连接在总线上的。为了将信息从一个部件传到另一个部件，源部件先将数据输出到总线上，然后目标部件再从总线上接收这些数据。随着计算机系统复杂性的不断增长，使用总线比每个设备对之间直接连接要有效得多（就减少连接数量而言）。与大量的直接连接相比，总线使用较少的电路板空间，耗能更少，并且在芯片或组成 CPU 的芯片组上需要较少的引脚。

数据是通过数据总线传送的。当 CPU 从存储器中读取数据时，它首先把存储器地址输出到地址总线上，然后存储器将数据输出到数据总线上，这样 CPU 就可以从数据总线上读取数据了。当 CPU 向存储器中写数据时，它首先将地址输出到地址总线上，然后把数据输出到数据总线上，这样存储器就可以从数据总线上读取数据并将它存储到正确的单元中。对 I/O 设备读写数据的过程与此类似。

控制总线与以上两种总线都不相同。地址总线由 n 根线构成，n 根线联合传送一个 n 位的地址值。类似地，数据总线的各条线合起来传输一个单独的多位值。相反，控制总线是单根控制信号的集合。这些信号用来指示数据是要读入 CPU 还是要从 CPU 写出，CPU 是要访问存储器还是要访问 I/O 设备，I/O 设备或者存储器是否已准备好传送数据等。控制总线实际是单向（大多数都是）信号的集合。大多数信号是从 CPU 输出到存储器与 I/O 子系统的，只有少数是从这些子系统输出到 CPU 的。

2.2 存储器

课文

任何一台数字计算机都需要存储 CPU 所执行的程序，因此，存储器是计算机最重要的部件之一。喜欢 PC 的人都爱谈论他们的存储器的容量有多大，但很少有人真正知道他们 PC 中的存储设备是怎样组成的。

1. 存储器的种类

RAM 和 ROM 在存储部件中扮演了重要的角色，我们应该了解它们。RAM 是随机访问存储器 3 个词首写字母的缩写。RAM 条主要用来做内存条。ROM 的意思是只读存储器。控制和专用程序存放在 ROM 芯片中，用户不能修改。

计算机带有厂家安装的，被称为只读存储器的永久性内存。基本操作指令集存储在 ROM 中，关闭计算机也不会被删除。过去，不换 ROM 模块或不换系统板就不能改变存储在 ROM 中的指令。计算机中有可以升级的模块，叫作 EEPROM（电可擦除可编程只读存储器）。BIOS（基本输入/输出系统）指令和配置应用程序都存储在计算机的快速 EEPROM 中。

除了永久性的内存外，计算机还有一个暂时性的内存。计算机工作期间，所得到的指令及所处理的信息都保存在 RAM 中。RAM 不是一个长久存储信息的地方。关闭计算机后，内存中就不再保存工作期间所输入的信息了。由于 RAM 只在开机时有效，所以计算机用磁盘驱动器来存储信息，这些信息在关机后依然存在。

计算机内存以信息的千字节或兆字节来度量（1 个字节等于一个字符、一个字母或数字的存储量）。1kB 等于 1024 字节，1MB 约等于 1000000 字节。软件需要一定数量的内存来正常工作。如要给计算机增加新的软件，在软件包装上通常可以找到该软件所需要的确切内存容量。

2. 存储器的单元

存储器由许多单元组成，每个单元可以存储一个信息。每个存储单元有一个号码，叫作单元地址。通过地址，程序可以访问存储单元。假定存储器有 n 个单元，它们就有地址编号 0～$n-1$。存储器的所有单元具有同样的位数。如果一个单元有 k 位，它可以保存 2^k 个不同位组合数据中的任一个。注意相邻的单元具有连续的地址。

使用二进制数字系统（包括使用二进制数的八进制和十六进制的记数法）的计算机，也用二进制表示存储器地址。如果地址有 m 位，可直接寻址的最大单元数量是 2^m 个。地址的位数与存储器可直接寻址的最大单元数量有关，而与每个单元的位数无关。具有 8 位长的 2^{12} 个单元的存储器和具有 64 位长的 2^{12} 个单元的存储器都需要 12 位地址。

单元的含义表示它是最小的可寻址单位。近年来，大多数计算机制造商已经使其长度标准化为 8 位，这样的单元叫作字节。字节可组成字，16 位字长的计算机每个字包含 2 个字节，而 32 位字长的计算机每个字则包含 4 个字节。字的含义是大多数指令对整字进行操作，比如把两个字相加在一起。因而 32 位机器有 32 位的寄存器和指令，以实现传送、加法、减法和其他 32 位字的操作。

在实模式下运行的 80×86 系列处理器，其物理寻址能力达到 1MB。EMS 采用页面调度或存储切换技术，使得微处理器能访问更大的存储空间。为了扩展存储器，需要额外的硬件和驱动程序。存储切换寄存器作为有 1MB 空间的物理窗口和驻留在扩展存储器上的逻辑存储器之间的接口。驱动程序，也称作扩展内存管理器（EMM），控制这个寄存器，使得程序的存储器访问在整个可用的扩展存储器上能实现重定向。

为了访问扩展存储器，程序需要与 EMM 联系。与 EMM 通信的方式与调用 DOS 类似。程序中设置合适的 CPU 寄存器值并建立软中断请求。定义了 30 多个功能，应用程序和操作系统可以控制扩展存储器。

当一个程序装入扩展存储器页中时，EMM 就将一个标志回复给这个请求程序。当再次调用 EMM 时，这个标志将用来区分逻辑页中哪些块被用过。

3．芯片的内部组成

ROM 和 RAM 芯片的内部组成是相似的。为了说明一个最简单的组成——线性组织方案，我们来考虑一个 64×4 的 ROM 芯片。这个芯片有 6 个地址输入端和 4 个数据输出端，以及 256 位的内部存储元件，它排列成 64 个单元，每个单元有 4 位。

6 个地址位经过译码，可以选择 64 个单元中的一个，但只有芯片的使能端有效才行。如果 CE=0，译码器被禁止，则不选择任何单元。单元上的三态缓冲器是有效的，允许数据输出到缓冲器中。如果 CE 和 OE 都设置为 1，则这些缓冲器有效，数据从芯片中输出；否则，输出是高阻态。

随着单元数量的增加，线性组织方案中地址译码器的规模变得相当大。为了解决这一问题，存储器芯片可以设计成使用多维译码方式。

在大型存储器芯片中，这种节省显得至关重要。考虑一个 4096×1 芯片，其线性组织方案将需要一个 12～4096 译码器，译码器大小与输出的数量成正比（因此一个 n～2^n 译码器的大小是 $O(2^n)$）。如果芯片排列成 64×64 的二维阵列，它将有两个 6～64 译码器：一个用来选择 64 行中的一行，另一个用来在选定行中选择 64 个单元中的一个单元，该译码器的大小正比于 2×64，或写成 $O(2 \times 2^{n/2}) = O(2^{n/2+1})$。对于这个芯片，两个译码器总的大小约是那个大译码器大小的 3%。

4．存储器子系统配置

构造包含一个简单芯片的存储器系统是非常容易的。我们只需要从系统总线上连接地址信号线、数据信号线和控制信号线就完成了。然而，大多数的存储器系统需要多个芯片，下面是通过存储器芯片组合来形成存储器子系统的一些方法。

两个或多个芯片可以组合起来构造一个每单元有多位的存储器。这可以通过连接芯片相应的地址信号线和控制信号线，并将它们的数据引脚连到数据总线的不同位上来完成（见图 2-3 略）。

例如，两个 64×4 芯片可以组合产生一个 64×8 存储器，如图 2-4 所示。两个芯片从总线上接收相同的六位地址输入，还有共同的芯片使能信号和输出使能信号（目前，我们只要了解两个芯片使用的是同样信号就可以了）。第一个芯片的数据引脚连到数据总线的第 7 位到第 4 位，第二个芯片的数据引脚则连在第 3 位到第 0 位。

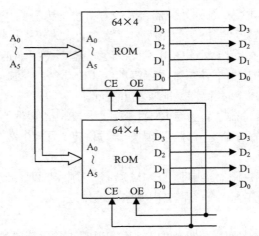

图 2-4　由两片 64×4 的 ROM 芯片组成 64×8 的存储器子系统

当 CPU 读取数据时，它将地址放在地址总线上。两个芯片读取地址位 A_5 到 A_0，并执行内部译码操作。如果 CE 和 OE 信号是有效的，两个芯片则输出数据到数据总线的八根线上。因为两个芯片的地址和使能信号是相同的，因此在任一时刻两个芯片要么同时有效，要么同时无效。计算机不可能只使一个有效。正因如此，它们的行为就像一片 64×8 的芯片一样，至少就 CPU 而言是这样的。

阅读资料

存储器访问

指令周期是微处理器完成一条指令处理的过程。首先，微处理器从存储器读取指令，然后将指令译码，辨明它取的是哪一条指令。最后，它完成必要的操作来执行指令。每一个功能——读取、译码和执行都包括一个或多个操作序列。

随着微处理器从存储器中取出指令，计算机开始运行。首先，微处理器将指令的地址放到地址总线上。存储器子系统输入地址并进行地址译码，寻址相关的存储单元。

当微处理器为存储器留出充足的时间来对地址译码和访问所需的存储单元之后，微处理器发出一个读（READ）控制信号。当微处理器准备好可以从存储器或是 I/O 设备读数据时，它就在控制总线上发一个读信号。一些处理器对于这个信号有不同的名字，但是所有的处理器都有完成这个功能的信号。根据微处理器的不同，读信号可能是高电平有效（信号=1），也可能是低电平有效（信号=0）。

读信号发出后，存储器子系统就把要读取的指令码放到计算机的数据总线上，微处理器就从数据总线上输入该数据并且将它存储在其内部的某个寄存器中。至此，微处理器已经取到了指令。

接下来，微处理器对这条指令译码。每一条指令可能要由不同的操作序列来执行。当微处理器对该指令译码时，它确定处理的是哪一条指令以便选择正确的操作序列去执行。这一步完全在微处理器内完成，不需要使用系统总线。

最后，微处理器执行该指令。指令不同，执行的操作序列也不同。执行过程可以是从存储器读取数据、写数据到存储器、读或写数据到 I/O 设备，执行 CPU 的内部操作，或者执行多个上述操作的组合。

微处理器从存储器读取数据所执行的操作序列，同从存储器中取一条指令是一样的。

符号 CLK 是计算机的系统时钟，微处理器用系统时钟使其操作同步。在一个时钟周期的开始位置，微处理器将地址放到总线上。一个时钟周期（允许存储器对地址译码和访问数据的时间）之后，微处理器才发出读信号。这使得存储器将数据放到数据总线上。在这个时钟周期之内，微处理器从系统总线上读取数据，并存储到它的某个寄存器中。在这个时钟周期结束时，微处理器撤销地址总线上的地址，并撤销读信号。然后存储器从数据总线上撤销数据，也就完成了存储器的读操作。

在存储器写操作的第一个时钟周期，处理器将地址和数据放到总线上，然后在第二个时钟周期开始时发出一个写（WRITE）控制信号。像读信号促使存储器读取数据一样，写信号促使存储器存储数据。在这个时钟周期的某个时刻，存储器将数据总线上的数据写入地址总线所指示的存储单元内。当这个时钟周期结束时，微处理器从系统总线上撤销地址、数据及写信号后，就完成了存储器的写操作。

存储器芯片的一个非常重要的特征参数就是数据的存取速度。为了存取数据，首先将地址信号传送到地址引脚上，一段时间以后，要存取的数据就会出现在数据引脚上。所用的时间越短，效果越好，存储芯片的价格也越高。存储器芯片的存取速度通常是指它的存取时间。

2.3 输入/输出设备

课文

计算机是功能强大的机器，它可以完成人们交给它的任何任务。然而计算机并不能够直接与人交流。借助于输入、输出设备，计算机和人便可以相互"了解"了。使用输入设备，人们告诉计算机它该做些什么，而计算机通过输出设备反馈结果。

输入/输出设备是人-机接口。它通常包括键盘、鼠标、显示器、打印机、磁盘（硬盘或软盘），输入笔、扫描仪和麦克风等。让我们来看看一些常用的输入、输出设备。

1. 键盘

键盘用来向计算机中键入或输入信息（见图2-5略）。键盘的布局和尺寸的大小有多种不同的设计方法，最常用的是以拉丁语为基础的 QWERTY 布局（以键盘上头6个键命名的）。标准键盘有 101 个键，笔记本电脑具有内置的键，可以通过专门的键或按键组合来访问。标准键盘上的某些键具有特殊的用途，有些键叫作命令键。最常用的3个命令键是控制键 Ctrl、替换键 Alt 以及轮换键 Shift。键盘上的每个键上面都有1~2个字符，按下一个键将得到下排的字符，按住 Shift 的同时将得到上排的字符。

数字键盘位于键盘的右边，看起来就像一个加法器。然而，当你使用数字键盘作为计算器时，要确信按下了数字锁定键（Num Lock），这样数字锁定键上方的指示灯会亮。

功能键（F1、F2等）通常位于键盘的上端，这些键用来给计算机发出命令，每个键的功能因软件不同而不同。

方向键允许你移动光标在屏幕上的位置。

专用键用来完成特定的功能。取消键（Esc）的功能取决于所用的程序，通常它将退出一个命令。屏幕打印键（Print Screen）会将屏幕上的信息发送到打印机输出。滚动锁定键（Scroll Lock）并不是在所有的程序中都可以操作，这个键在现在的软件中很少使用。数字锁定键控制数字键盘的使用。大写字母锁定键（Caps Lock）可以控制打出的文本都是大写字母。

2. 鼠标

鼠标是一种小型的设备。用户可以在桌面上移动，以指向屏幕上的某个位置并从该位置选择一项或多项操作（见图2-6略）。最常见的鼠标顶部有两个按键。左键用得最多。在Windows操作系统下，用户可以单击此键，发出一个"选中"信号，继而得到来自系统的反馈，表示指定的位置已经选中，并可进一步操作。在选中的位置再按一次该键或双击此处，就可以对选中的对象进行某种操作。第二个按键在右边，提供了某些不太常用的功能。例如，在浏览网页时，用鼠标右键单击图像可以弹出一个菜单，其中有一项命令可以将图像存盘。

鼠标由以下几个部分组成：一个金属或塑料的盒体，一个凸出于盒体底部并可以在平面上滚动的球体，位于盒体上部的一个或多个按键，以及一条连接到计算机的电缆线。球体在平面上沿任意方向滚动时，鼠标内部的传感器将相应的脉冲信号传输到计算机，这时支持鼠标操作的程序随即做出响应，将可视化的指示器（光标）在屏幕重新定位。光标的定位相对于其开始位置。看到光标的当前位置后，用户可以移动鼠标再次进行调整。

3．显示器

当你打字时，显示器会在屏幕上显示信息，这叫作输出信息。当计算机需要更多的信息时，它将在屏幕上显示一条信息，通常是通过一个对话框。显示器有多种类型和尺寸，从简单的黑白（单色）显示器到全彩色显示器都有（见图2-7略）。

基于字符的显示器将屏幕分成矩形网格，每个网格可以显示一个字符。显示器可显示的字符集是不可修改的，因此要显示不同大小和格式的字符是不可能的。位图显示器将屏幕划分成微小的矩阵，称为像素的矩阵点。在计算机屏幕上显示的任何字符或图形必须由屏幕矩阵的点阵格式构成。屏幕在矩阵中显示的点越多，分辨率就越高。

- 分辨率：分辨率指的是显示器所能容纳的单个色点数，即像素数。通常分辨率是由水平方向（行）的像素数和垂直方向（列）的像素数来确定，如640×480。显示器的可视区域、刷新率以及点距对最大分辨率都有直接的影响。
- 点距：简单地说，点距是对显示器像素之间空间大小的量度。当考虑点距时，记住点距越小越好。要获得更高的分辨率，重要的是使像素之间离得更近。显示器通常能支持与点（像素）物理尺寸大小相匹配的分辨率，也能支持一些较低的分辨率。
- 刷新率：在基于CRT技术的显示器中，刷新率是指显示器上的图像每秒钟绘图的次数。刷新率是非常重要的，因为它控制闪烁，刷新率越高越好。

4．打印机

常见的打印机是点阵打印机。便宜的打印机可能有7根针，用5×7的阵列在每行打印80个字符。实际上打印行由7个水平行组成，每行有5×80=400个点。每个点可能打印或不打印，这取决于所打印的字符。通过两种技术可以提高打印质量：使用更多的针和循环重叠。

喷墨打印机使用喷嘴把墨水喷射到纸上以形成适当的字符。为了得到正确的字符，墨水受到一个阀门和一个或多个控制喷射流的垂直和水平位置的偏转器的控制。可以打印大量的具有不同风格的不同字符。某些喷墨打印机具有打印全彩色图像的能力。

激光打印机使用激光束照射激光感光纸。于是，这种纸得到粉粒或增色剂，并以热量、压力或二者结合的方式使粉粒或增色剂固定在纸上。使用激光打印机，一次可以打印一整页。激光打印机的缺点之一是静电问题。在这些打印机中，打印纸可能有粘在一起的问题，这使得用激光打印机为顾客设计票据和财务清单变得很困难。为避免这个问题，已开发出了新型的冷激光打印机。它们不需要热量就可以把字符固定在纸上。这样，静电和粘纸问题可以消除或大幅度减少。

5. 调制解调器

为了在远距离的计算机间通信，人们想到了利用现有的电话网来进行数据传输。大多数的电话线是用来传输模拟信号（声音）的，然而计算机及其设备却是以数字形式（脉冲）工作的。因此，为了利用传输模拟信号的媒介，在两种系统之间需要一个转换器。这个转换器就是调制解调器。调制解调器可以将数据从数字信号转换成可以通过电话线传输的模拟信号，这个过程叫作调制。在接收端，又将模拟信号转换为数字信号，这个过程叫作解调。

调制解调器可以分为外置式和内置式。外置式调制解调器面板上有一系列信号灯，若解决安装时遇到问题，信号灯可以提供重要的信息。在安装时，需要用电缆将调制解调器正确地连接到计算机未占用的串行端口上。内置调制解调器是一块印刷电路板，将内置调制解调器安装在计算机中将占据一个扩展槽。

6. 其他设备

随着输入/输出设备技术的发展，近年来扫描仪也渐渐走进了普通家庭。它能以最便捷的方式，对印在纸上的各种信息进行输入，而且借助于识别和分析软件，一切只需几分钟就可以扫描到计算机中。

如果软件带有音乐，你就需要一个音箱连接到计算机上来播放。你可以在计算机上边工作边听音乐，那可真是一种享受。你曾经试过用你的声音来控制计算机吗？借助于麦克风，语音控制已经在一些非常流行的文字处理软件中使用了。

阅读资料

组装计算机

自己组装计算机不仅仅是一项需要掌握的很有价值的技能，而且这项技能也是非常简单、易学易懂的。

1. 安装 CPU 和 RAM

主板在机箱外时，安装 CPU 就容易多了，因为这样可以有更多的自由工作空间。首先，要确信自己的 CPU 是带散热片或风扇的。接下来，将 CPU 插入 CPU 插槽中。尽管取决于安装的 CPU 的类型，但安装 CPU 正确的方法只有一个，轻轻地把 CPU 插到插槽上，然后检查并确信已经插牢了。

在主板上应该能找到 RAM 插槽。现在大多数主板都有两个 DIMM（双列直插内存模块）连接器，安装 RAM 内存条是件轻而易举的事。如果使用的是 DIMM，则内存条的尺寸、安装的位置和顺序关系不大。最好把第一条安装到 bank 0 上（通常这个位置最靠近 CPU），而且只需要将内存条的金色触点插座对准合适的插槽，然后适当用力，但不要太用力。

确信所有的跳线都正确设置并且 CPU 和 RAM 也已经安装好了。现在需要插好机箱 LED 指示灯（电源和硬盘指示灯）、电源开关和复位开关等接线。每块主板的机箱接线安排的都不同，所以参考主板手册以便正确地排列这些线。

2. 安装 IDE 设备

现在，大多数主板都可以支持多达 4 个 IDE（集成设备电子）设备。需要做的第一件事情就是设置 IDE 设备跳线。可以找到三种方法来设置这些跳线：主设备、从设备或电缆选择。最好是将硬盘（图 2-8 略）设置成第一个 IDE 口的主设备，然后将 CD-ROM 或 DVD-ROM 驱动器作为第二个 IDE 口的主设备。记住每一个 IDE 口都只能有一个主设备和一个从设备。

3．插入 AGP/PCI 设备

只需要将卡插入 AGP 槽（通常是离 CPU 最近的褐色插槽），并确信已经插牢，然后用螺丝刀将插座的边固定到机箱上。现在可以试着启动一下计算机了。如果一切正常，也即计算机通过自检，可以关掉电源然后继续安装 PCI 卡，比如声卡、调制解调器和其他需要安装的部件。最好将各种卡隔开以便空气流通。完成这项工作后，准备第一次真正的启动系统了。

4．设置 BIOS

当计算机成功地加电自检查后就需要设置 BIOS（基本输入/输出系统）了。在计算机自检后按 Delete 或 F2 键就可以进入计算机的 BIOS 设置界面了。翻阅你的主板手册就可以访问并设置 BIOS 了。

5．安装操作系统

到目前为止，最容易安装的操作系统就是 Microsoft Windows。安装操作系统无非就是使用软驱或光驱来启动计算机，并在安装时按照简单的屏幕指令去做。

安装新版本的 Windows 系统时，可以将旧版本的 Windows 系统保留在计算机上。计算机上安装了多个操作系统的情况称作"多重引导配置"。开始安装之前，请确保硬盘针对每个要安装的操作系统都有一个单独的分区，或计算机有多个硬盘。否则，必须对硬盘进行重新格式化和重新分区，或将新的操作系统安装在不同的硬盘上。此外，请确保计划安装新版本 Windows 的分区或磁盘采用 NTFS 文件系统进行格式化。如果计算机上安装了多个操作系统，则可以在打开计算机时选择启动哪个操作系统。

一般来说，软件安装更为简单。当然，如果希望更为熟练地掌握安装的基本技巧，必须进行更多的实践。当使用计算机时，必须知道如何与它"交流"。在一些特殊情况下，计算机会发出一些有用的出错信息或提示来帮助你继续工作。下面是一些最常见的信息。

（1）错误的文件名或命令。
（2）终止、重试或取消？
（3）请阅读下列授权协议书。
（4）建议安装前先关闭其他所有的应用程序。
（5）请单击 OK 按钮继续安装。
（6）按 F1 键请求帮助。
（7）重启计算机，完成安装。

第 3 章 软件知识

3.1 操作系统

课文

操作系统究竟是做什么的呢？从根本上讲，它完成了大量的支持功能。例如，描述存储在磁盘上的应用程序，因为控制计算机的程序必须在主存中，所以存储在磁盘上的应用程序在执行之前必须被复制到主存中。把程序从磁盘复制到主存的过程包含了必不可少的逻辑处理，计算机的逻辑处理的来源就是软件。因此，如果要装入应用程序，则必须在主存中有一个控制加载过程的程序，这个程序就是操作系统。

操作系统有单任务和多任务之分。早期的许多单任务操作系统同一时间只能运行一个程序。例如，当计算机打印文件时，它就不能开始运行另一个程序，或者不能响应新的命令，直到打印完成为止。

所有现代操作系统都是多任务的，同时能运行几个程序。大部分计算机中仅有一个 CPU，所以，多任务操作系统让人产生 CPU 能同时运行几个程序的错觉。用于产生这种现象的最常用机制是时间片多任务处理，以每个程序各自运行固定的一段时间的方式来实现。如果一个程序在分派的时间内没有完成，它就被挂起，另一个程序接着运行。这种程序交换称为任务切换。操作系统实行"簿记"法以保存被挂起的程序状态。它同样有一种机制，叫作调度程序，由它决定下一时刻将运行哪个程序。为了把感觉到的延迟减到最小，调度程序运行短程序非常迅速。由于用户的时间感觉比计算机的处理速度要慢得多，所以几个程序看起来是同时执行的。

当实际可用空间不够时，为了运行那些需要更多主存储器空间的程序，操作系统可以利用虚拟存储器。采用这一技术，硬盘空间被用来模拟所需的额外存储空间。不过访问硬盘比访问内存更加耗时，所以计算机性能会下降。

1．资源分配及相关功能

资源分配功能分配资源供用户计算使用。资源可分为系统提供的资源，如 CPU、存储区域和 I/O 设备，或用户创建的资源，如由操作系统管理的文件等。

资源分配的标准根据资源是系统资源还是用户创建的资源来决定。系统资源的分配要考虑资源利用的效率，而用户创立资源的分配则基于该资源的创立者所设定的特种限制，比如访问权限。

资源分配通常采用以下两种策略：
- 资源分区；
- 从资源池中分配。

在资源分区方式中，操作系统预先决定把哪些资源分配给某个用户计算机使用，这种方法也称为静态分配，因为分配是在程序执行前进行的。静态资源分配易于实现，但由于它不是从程序的实际需要出发，而是根据程序预先提出的需求来做决定，所以容易导致系统利用率下降。在后一种分配方式中，操作系统维护一个公共资源池，并按照程序的需要对资源进行分配。这样，操作系统只在程序提出对一个资源的需求时才进行资源分配，这种方式也称为动态分配，因为分配是在程序执行的过程中进行。动态存储分配的资源利用率较高，因为它是在程序需要资源时才进行分配。

操作系统可以利用资源表作为资源分配的中心数据结构。系统的每个资源在表中都有一个数据项来表示。数据项中记录资源单位的名称或地址以及当前状态，即它是空闲的还是已经分配给了某一程序。当程序对某一资源提出请求后，若该资源是空闲的，则它将被分配给那个程序。若系统中同一资源类中存在许多资源单位，程序的资源请求只指明要求哪类资源，而由操作系统检查该类中是否有可用的资源单位可以分配。

CPU 只能串行共享，因此它一次只能执行一个程序，其他程序必须依次等待。通常情况下，系统会平等对待所有的程序，因此用抢占来释放 CPU 以执行其他程序。决定该执行哪个程序并执行多长时间是一个十分重要的功能，这一功能被称为 CPU 调度或简称调度。显然，资源分区不适用于 CPU 共享，因此，从资源池中分配成为唯一的选择。

和 CPU 一样，存储器也不能并行共享。但与 CPU 不同的是，可以把存储器的不同部分看作不同的资源，因此可以增加它的可用性。资源分区和基于资源池的分配方式都适用于存储器资源管理。

2．I/O 操作控制

系统资源的分配与操作系统软件对 I/O 操作的控制密切相关。由于 I/O 操作开始之前需要对指定设备进行访问，因此操作系统必须协调 I/O 操作和使用设备间的关系。实际上操作系统建立了一个执行程序和完成 I/O 操作必须使用的设备的目录。使用控制语句，作业可以访问指定设备。因而用户可以在指定设备上读取数据或在选定的办公室打印信息。利用这一功能的优势，读自某一设备的数据可以分布贯穿于整个计算机系统。

为便于 I/O 操作的执行，大多数操作系统都有一个标准的控制指令集来处理所有输入和输出指令。这些标准指令称为输入/输出控制系统（IOCS），是大多数操作系统不可分割的部分。它们简化了被处理的程序承担的 I/O 操作。

实际上，使用一个特殊的 I/O 设备时，程序在执行的过程中向操作系统示意所要求的 I/O 操作。控制软件访问 IOCS 软件以实际完成 I/O 操作。由于大多数程序都考虑 I/O 操作的级别，所以 IOCS 指令至关重要。

3．操作系统结构

操作系统的设计主要基于以下两个因素：它所操作的计算机的结构特征以及它的应用控制特征。对结构特征的依赖是由对系统的所有功能单元进行完全控制的需要而产生的。因此，操作系统需要了解计算机系统的地址结构、中断结构、I/O 体系结构及存储器保护特性。操作系统的策略通常由它的应用控制决定。例如，CPU 的调度策略是由操作系统是用于分时应用还是实时应用来决定的。对这两方面因素的依赖导致在不同结构和不同应用控制中的计算机上使用同一操作系统的巨大困难。

考虑设计一个操作系统，使它能应用于处在相似应用控制下的两个计算机系统 C1 和 C2 上。两个操作系统除了与结构相关的操作系统代码有所不同外，其余大部分操作系统代码都与结构无关。则可以考虑采用以下方式开发用于 C1 和 C2 的操作系统。

- 设计用于计算机系统 C1 的操作系统，设其为 OS1。
- 把 OS1 修改为能在计算机系统 C2 上使用的操作系统，即 OS2。

也就是说，OS2 是通过把 OS1 移植到 C2 上得到的。

由于早期的操作系统均是整体结构，就给这种方式的实施造成了困难。因为操作系统并没有为结构相关的及结构无关的代码之间提供清晰的接口，因此移植工作就要涉及全部的操作系统代码，而不是其中只和结构相关的部分。历史上，这一困难是通过设计一种把结构相关的部分和结构无关的部分分开的操作系统结构来克服的。这样可以使不同计算机系统的操作系统分享大部分的设计。若操作系统是用高级语言编写的，则不同的操作系统之间甚至可以共享代码。

阅读资料

Linux 操作系统

Linux 是一种操作系统，原来是由芬兰赫尔辛基大学的年轻学生 Linus Torvalds 作为业余爱好编制出来的。1991 年，他发布了 0.02 版本并由此开始了他的工作。直到 1994 年 Linux 内核 1.0 版本被发布，他的工作才稳定了。这个内核，位于所有 Linux 系统的中心，是在 GNU 通用公共授权下开发和发布的，它的源代码是开放使用的。正是这一内核成为 Linux 操作系统开发的基础。目前有上百家公司和机构以及同样数量的个人基于 Linux 内核发布了他们自己的操作系统版本。

Linux 具体体现了 Unix 的设计思想，这就意味着它是多用户系统。它也有许多优点。无论是一个用户正在运行几个程序，还是几个用户正在运行一个程序，Linux 都能管理好信息传输。

　　Linux 是开放的。这意味着所有的专业编程人员和用户不仅有权修改它，也可以访问源代码。这对用户来说有许多好处。它意味着软件的质量更好、运行的效率更高和系统崩溃的可能性更小。人们也更容易参与其开发过程，也就是说，即使不是专业编程人员的人，他也可以通过向开发小组提出改进建议，而对一个软件给予极大的影响。

　　Linux 有几种形式。从技术方面而言，有简单的内核源代码、各种程序以及自己可以把它们组装在一起的应用。但是，对那些不想一点一点拼凑自己系统的人来说，Linux 有许多发布。这些发布在设置方法和为软件打包的方法上略有不同。它们在为用户提供某些媒体方面是类似的，通常是一组 CD 或一组软盘，通过它们就可以安装 Linux 了。它们也提供图形工具来安装和配置系统。

　　然而，Linux 的一个主要问题就是安装。它不像安装其他操作系统那样顺畅。从某种意义上说，用户可能不得不学会简明的 Unix 命令行语句。这一原因部分来自于 Linux 是一个配置性极强的系统，部分来自于对 Unix 的继承。

　　为了更好地理解 Linux 系统的工作方式，或如何定制该系统以适应自己的需求，应该阅读一下 HOWTO 文件。这些文件解释了如何在 Linux 系统上做任何事情，从选择硬件到安装设置等都做了说明。HOWTO 文件在许多 Linux 资源网站上都能找到。

　　Linux 是现有的较为稳定的操作系统之一。这主要因为专业编程人员是在为其他专业编程人员编写，而不是为公司团体编写。唯一决定要把什么编写进系统的人是专业编程人员。而且，当一个人把它作为业余爱好来开发时，软件期限的压力就不会那么大。

　　Linux 系统对大多数应用来说性能良好。但是在网络负载很重的情况下表现不佳。在同样的硬件环境下，Linux 的网络性能要比 FreeBSD 系统低 20%~30%。Linux 对此问题已有所改进，在 2.4 版本内核中，它将采用一种新的虚拟内存技术，这种技术类似于 FreeBSD VM。由于 Linux 和 FreeBSD 都开放源代码，因此好的技术可以在两种系统中共享，也正是由于此原因，两种系统在很多方面是相似的。

　　安装 Linux 系统的服务器可以长年稳定地工作。但是 Linux 系统的磁盘 I/O 是非同步的，这降低了基于操作事务的稳定性，在系统出错或者电源失效的情况下会导致文件系统的错误。但是对一般用户而言，Linux 系统已经非常可靠了。

3.2　数据结构

课文

1．数据结构介绍

　　数据以各种形态和大小出现，但是它常常可以用同样的方式来组织。例如，考虑要做事情的列表、处方成分的清单或一个班级的阅读目录，虽然它们包含不同类型的数据，但它们都包含以一种相似方式组织的数据：一个列表。列表是数据结构的一个简单例子。当然，还有许多其他组织数据的通用方法。在计算机技术中，一些最常用的组织方式是链表、堆栈、队列、集合、哈希表、树、堆、优先队列和图。效率、抽象和复用性是使用数据结构的主要原因。

　　数据结构使用令算法更有效率的方法组织数据。例如，考虑一些我们用来查找数据的组织方式。一种极为简单的方式是将数据放置到数组中，并用遍历的方法找到需要的元素。然而，这种方法是低效率的，因为在许多情况下，我们需要找遍所有元素才能完成。使用其他类型的数据结构，如哈希表和二叉树，就能够相当快速地搜寻数据。

数据结构提供一个更好理解的方法查找数据，因此，它们在解决问题中提供一定的抽象化水平。例如，通过把数据存储在堆栈中，我们可以将重点集中在对堆栈的操作上，如使元素进栈和出栈，而不是集中在实现操作的细节上。换句话说，数据结构使我们以较少的编程方式谈论程序。

因为数据结构趋向于模块化并和环境无关，所以数据结构是可以复用的。因为每种结构有一个预定的接口，通过该接口限制访问存储在数据结构中的数据，所以他们是模块化的。也就是说，我们只能使用接口定义的那些操作来访问数据。因为数据结构能用于任何类型的数据，并用于多种环境中，所以数据结构与使用环境无关。在C语言中，我们通过使用空指针，而不是通过维护非公开的数据备份，使数据结构存储任何类型的数据。

面向对象的程序设计是一种现代的软件开发方法，用这种方法设计的软件具有高的可靠性和灵活性。在特定类中，从决定使用何种数据结构来表示对象的属性这一点来看，面向对象方法中对抽象的强调，在软件开发过程中是非常重要的。抽象意味着隐藏不必要的细节。

过程抽象或算法抽象是对算法隐藏细节的，允许算法在各个细节层次上可见或被描述。建立子程序是抽象的一个实例，子程序名描述了子程序的功能，子程序内部的代码表示了处理过程是如何完成的。类似地，数据抽象隐藏了描述的细节。一个明显的例子是通过把几种数据类型组合来构建新的数据类型，每种新类型描述了一些更复杂的对象类型的属性或组成。数据结构中面向对象的方法通过把一类对象的表示整合将数据抽象和过程抽象组合在一起。

2．数据结构与算法

数据结构是用来组织和存储数据的一种特殊形式。一般的数据结构类型包括数组、文件、记录、表、树等。任何数据结构都是用来组织数据以适应某种特定的目标，从而以适当的方法达到存取和操作的目的。在计算机程序设计中，可能会选择或指定一种数据结构来存储数据，以便在这种结构上实现各种算法。

在现实世界中对问题的描述可能会有许多不必要的细节。解决问题最重要的一步就是要识别基本的抽象问题，而去除不必要的细节。类似地，一个具体的计算机模型对一个问题也有许多不相关的细节，比如处理器的结构和字长等。计算机程序设计的艺术之一就是消除一个问题中和所用计算机不必要的细节。

一旦问题被抽象化以后，看起来似乎是不同的，但在深层意义下很明显本质上是类似的甚至是相同的问题。例如，维护一个学生的听课名单与组织一个编译程序词典结构有很多相同的地方；二者都要对确定的事物进行存储和操作，而这些事物又都有特定的属性或性质。抽象概念允许对看似不同的问题具有相同的解法。利用抽象算法和数据结构求解问题具有最大的实用性和被重新使用的机会。

算法和数据结构可以用任何足够准确的语言来描述。如果小心使用，避免多义性，英语和其他自然语言都是可以用的，但一般最好是使用更为准确的数学语言和程序设计语言。后者执行起来也能够自动实现。程序是用某种特定的程序设计语言对某种算法和数据结构的表达。程序可以在带有相应编译程序或解释程序的任何类型计算机上使用，这使得程序更具有价值。

对给定的求解问题，我们要选择或设计抽象数据结构以存储数据和算法，并对它们进行操作。数据结构是数据类型的实例。其设计步骤独立于任何程序设计语言。其编码或设计步骤就是用特定的程序设计语言，如图灵语言，对其抽象数据结构和算法程序的实现。抽象数据类型要定义数据的某种静态性质以及定义数据上的某种动态操作。它定义所操作的内容是什么，而不是如何操作。而算法则定义一些过程以及一个操作是如何实现的。它是关于一些简单步骤的规定。每个步骤必须能用一种机械方式明确地执行，它也许是一条单个机器指令、一条程序语言的语句或者是前面定义过的一个算法；一个求解问题的算法必须是正确的；当给定有效的输入时，它必须能产生正确的输出；当给定有效的输入时，一个算法必须能够终止。当输入的数据无效时，我们通常希望算法能给出我们警告并能够终止。我们通常希望知道一个算法要耗费多长时间或它要使用多大空间（计算机内存），这就引出了算法分析和计算复杂性的问题。

3．数据类型

数据类型的本质是标识一组个体或目标所共有的特性，这些特性把该组个体作为可识别的种类。如果我们提供了一组可能的数据值以及作用在这些数据值上的一组操作，那么，这两者结合在一起就称之为数据类型。

来看两种数据类型，我们称任何由原子值构成的数据类型为原子数据类型，通常我们倾向于把整数作为原子。那么，我们关心的仅仅是一个值所代表的单个量，而不是把整数看成是一个在某些数字系统中的数字的集合。在许多程序设计语言和计算机体系结构中的整数类型是一个常用的原子数据类型。

人们可以将其值由某种结构相关的组成元素构成的数据类型称之为结构化数据类型或数据结构。换句话说，这些数据类型的值是可分解的，因此我们必须知道它的内部结构。任何可分解的目标有两个必要的组成成分——必须具有组成元素和结构，亦即将这些元素相互关联或匹配的规则。

（1）数据类型的分类：
- 原子数据类型（值不可分解）。
- 数据结构（值可分解）。

（2）数据结构——一种数据类型，其值：
- 能被分解成一组成员数据元素，每个数据元素或者是原子，或者是另一种数据结构。
- 包括一组涉及组成元素的联系或关系（结构）。

数据结构是一种数据类型，其值是由与某些结构有关的组成元素所构成的。它有一组在其值上的操作。此外，可能有一些操作是定义在其组成元素上的。由此我们可知：数据结构可以有定义在构成它的值之上的操作，也可以有定义在这些值的组成元素之上的操作。

阅读资料

栈和队列

在堆栈中，插入和删除操作都在结构的相同末端进行。如此限制的结果就是最后一个进入表的记录也就是第一个从表中删除的记录。这种结构称为后进先出（LIFO）结构。

堆栈尾部可以进行插入和删除操作的记录称为堆栈的栈顶，另一端叫作栈底。为了表示如何限制堆栈只能从栈顶访问，我们用特殊的术语来表示插入和删除操作。把一个对象插入堆栈的过程称为进栈操作，而从堆栈中删除一个对象的过程称为出栈操作，所以我们常说将一个记录进栈或者将其出栈。

为了在计算机存储中实现栈结构，一般采取的方法是保留一块足够容纳栈大小变化的内存单元。通常来说，确定块的大小是一个很重要的任务。如果保留的空间过小，那么栈最后可能从所分配的存储空间中溢出；而如果保留的空间过大，又是一种浪费。块的一端作为栈底，栈的第一条数据会被存储在这里，以后的条目被依次放置在它之后的存储单元中，也就是堆栈向另外一端增加。

因此，在条目进栈和出栈的时候，栈顶的位置就在存储单元块中前后移动。为了保存这个位置的信息，栈顶条目的地址被存储在一个附加的存储单元中，这个附加的存储单元称为堆栈指针。也就是说，堆栈指针就是一个指向栈顶的指针。

队列的例子在日常生活中经常出现并且为我们所熟悉。队列的主要特征是遵循先来先服务的原则。

队列的先进先出（FIFO）原则在计算机中有很多应用，例如，在多用户分时操作系统中，多个等待访问磁盘驱动器的输入/输出（I/O）请求就可以是一个队列。等待在计算机系统中运行的作业也同样形成一个队列，计算机将按照作业和 I/O 请求的先后次序进行服务，也就是按先进先出的次序。

用来在计算机中控制队列的最一般的方法是为队列分配一块存储器，从存储块的一端开始存储队列，并且让队列向另一端增长。当队尾到达块的末端，我们开始将新的条目反向于末端的方向插入，此时它是空闲的。同样，当队列的最后一条成为队头并被移出时，调整头指针回到块的开端，同时在此等待。在此方法下，队列在一块区域内循环而不会出现内存溢出情况。

采用此技术的实现方法称为循环队列，因为分配给队列的一块存储单元组成了一个环。就一个队列而言，存储块的最后一个单元与它的第一个单元相邻。

3.3 编程语言

课文

每一种程序设计语言都有一套专用的词汇及语法规则，用以指示计算机执行某种任务。广义地说，程序设计语言包括一套人和机器都可以理解的语句或表达式。人可以理解，是因为这些语言使用的是人们常用的短句（英语的或数学的）；另一方面，计算机借助于一种称作翻译器的专用程序，对上述指令进行处理。翻译器将人输入的指令予以解码，并生成机器语言代码。

（1）机器语言

能被计算机操作系统直接运行的计算机程序称为可执行程序。可执行程序是以机器代码的形式表示的一系列非常简单的指令。这些指令对于不同计算机的 CPU 而言是特定的，它们与硬件有关。例如，英特尔"奔腾"和 Power PC 微处理器芯片各自有不同的机器语言，要求用不同的代码集来完成相同的任务。机器码指令数量很少（20~200 条，根据计算机和 CPU 的不同而有差异）。典型的指令是从存储单元取数据，或将两个存储单元的内容相加（通常在 CPU 的寄存器中进行）。机器码指令是二进制的——也就是比特序列（0 和 1）。由于这些数字令人难以理解，所以计算机指令通常不是用机器码来写的。

（2）汇编语言

位于机器语言和高级语言之间的是汇编语言，这种语言与计算机的机器语言直接相关；换句话说，每一条汇编语言指令仅对应于一条机器语言指令。机器语言全部由数字组成，人们要读写几乎不太可能。与机器语言相比，汇编语言具有同样的结构和指令集，只不过允许程序员使用名称，以避免直接使用数字。因此，汇编语言所使用的命令较容易为程序员所理解。

一旦汇编语言程序编写完毕，它就由另一个称之为汇编器的程序转换成机器语言程序。由于汇编语言与机器语言保持一致性，因此速度快，功能强。可它仍然难以利用，因为汇编语言指令是一系列抽象代码。另外，不同的 CPU 使用不同的机器语言，因此需要不同的汇编语言。有时为了执行特殊的硬件任务，或者为了加快高级语言程序的速度，汇编语言被插入到高级语言程序中。

（3）高级语言

高级程序语言是这样一种编程手段，它用规范化的术语来写出一步步的程序步骤，执行这些步骤时会用唯一确定的方式处理给定的数据集。高级语言与任何给定的计算机无关，但假定该计算机是可用的。

高级语言经常针对一类特殊的处理问题而设计。例如，一些语言设计来处理科学计算类型的问题，另一些语言则似乎侧重文件处理的应用。

我们希望高级语言能便于做下列事情：
- 构造程序；
- 定义数据元素，赋以名字，字长和类型；
- 处理数据元素（算术/布尔/转换）；
- 控制程序流（测试、分支）；
- 允许通用程序重复利用（循环、子程序、过程）；
- 允许数据的输入/输出。

（4）C++和面向对象编程

C++是在 C 语言的基础上开发的，是 C 的新版本。它是由贝尔实验室的 Ejarne Stroustrup 开发的一种通用的、更易理解的应用程序设计语言。C++保留了 C 语言的大量特性，包括丰富的算符集，接近正交设计，简明扼要和具有可扩充性。C++是一种高度可移植性语言。它的编译程序存在于很多不同的计算机和系统中。由于保持这种可兼容性是一种设计目标，C++的编译程序与现在的 C 程序是高度兼容的。不像其他的面向对象语言，如 Smalltalk 语言，C++是对广泛应用于多种机器的现有语言的一种扩展。

C++是低级语言与高级语言的一种密切结合。C 语言原来是被设计用作一种系统实验语言，即接近机器的一种语言。C++增加了面向对象的特性。这些特性使程序员可以建立或导入适用于问题域的库。当仍与机器层次的实现细节保持联系时，用户可以在适合于问题的层次上编写代码。

C++完全支持面向对象编程，它包括 4 个面向对象的开发工具，即封装、数据隐藏、继承和多态。

（1）封装和数据隐藏

当一个工程师需要给自己正建造的设备加一个电阻时，通常他不会从头开始去制作一个电阻。他会到电阻库中检查表示属性的彩色条纹，选择一个满足其要求的电阻。对这位工程师来说，电阻是一个"黑盒子"——只要电阻满足他的规格即可，他不必关心电阻的内部结构及其工作原理。

能够成为自包容单元的属性称为封装。利用封装我们就可以实现数据隐藏。数据隐藏是一种非常有价值的特性，用户不必了解或关心某个对象的内部情况就可以使用它。就像你可以使用冰箱而不必关心其压缩机的工作原理一样，你也可以使用一个设计优良的对象而不必了解其内部的数据成员。

同样，工程师在使用电阻时也不必关心电阻的内部状况。电阻的所有属性都封装在电阻对象中；它们不会通过电路传出。不必去弄清楚电阻是如何工作的，我们就可以有效地使用电阻。其数据隐藏在电阻壳内部。

C++通过创建称为类的用户定义类型而支持封装和数据隐藏的属性。类一经创建，就可用作一个完全封装的实体——一个完整的单元。类的实际内部工作都应该被隐藏。一个定义好类的用户只需知道怎样使用它就可以了，而不必知道类是如何工作的。

（2）继承和重用

当 Acme 汽车工程师要造一辆新车时，他们有两种选择：或者从头做起，或者修改一个现有车型。也许他们的 Star 型车已接近完美，但他们想加一个涡轮增压器和一个六速变速装置。总工程师不想从头开始，于是说："让我们来另造一台 Star 车吧，但是，我们要给它增加这些附加装置。我们就把新车型叫作 Quasar 吧"。其实 Quasar 只是一种增加了一些新特性的 Star 型车。C++通过继承支持重用的思想。可以声明一个新类型作为已有类型的扩展。这个新的子类从已存在类派生而来，可称为派生类。Quasar 是从 Star 派生而来，因而它继承了 Star 的所有品质，只是根据需要增加了一些新的功能。

（3）多态性

对于新的 Quasar 型车，当你踩下油门时，它可能做出与 Star 型车不同的反应。Quasar 可能喷射燃料并启动涡轮增压器，而 Star 只是让汽油进入化油器中。但对用户而言，他不必知道这些差别。他只需踩下油门，根据他所驾驶的是哪种车，正确的动作随之发生。C++支持这种想法，即通过所谓函数多态性和类多态性不同的对象做"合适的事情"。"poly"意味着多，"morph"意味着形式，多态性是指同一名称的函数和类有多种形式。

阅读资料

软件工程

软件工程是应用各种工具、方法和规律来产生和维护现实世界问题的自动解决方案。它需要识别出问题、能执行软件产品的计算机以及软件产品存在的环境（由人员、设备、计算机、文档等组成）。

软件工程是在给定的资源条件下制作出满足用户需求的、可行的、高质量软件的学科。它是关于人们如何从事工作和利用技术来制作和维护软件产品和软件集成系统过程的方法。有关的概念包括规格说明、设计、实现、测试、验证和软件产品的评价等。相关的论题还有软件规格、项目管理及过程完善。软件工程对从事开发、维护、管理或与软件机构相关领域的任何人而言都是极为重要的。

软件工程的总体目标是使读者领悟一个集成系统和软件开发过程中事件的流程，体会并理解软件工程师在系统开发过程中的角色，为读者将来承担软件工程师的职责做好综合性的准备。

软件工程的早期方法坚持要严格地遵守分析、设计、实现以及测试的顺序。在大型软件系统的开发过程中，感觉到采用试错法是在冒着很大的风险进行开发的。因此，软件工程师坚持应当在设计之前进行完整的系统分析，同样，设计应该在实现之前完成。这就形成了一个现在称为瀑布模型的开发过程，这是对开发过程只允许以一个方向进行的事实的模拟。

近年来，软件工程技术已经开始反映这种本质的对立了，这可以由软件开发中出现的增量式模型来说明。根据这个模型，所需的软件系统是通过增量模式来构造的——首先开发最终产品的简化版本，它只有有限的功能。一旦这个版本经过测试，并且也许经过了未来用户的评估，更多的特性就可以逐渐地添加进去并且进行测试，直到完成系统。例如，如果正在开发的系统是为大学注册人员设计的学生记录系统，第一次版本仅仅包括浏览学生记录的功能。一旦这个版本可以运行了，其他特性，诸如增加和更新记录的功能，就可以分阶段地添加到系统中了。

这种增量式模型是软件开发向原型法发展趋势的一个证据——在原型法中，建立并测试的是不完善系统，被称为原型。在增量式模型中，这些原型进化为一个完整的最终系统，这个过程称为演化式原型。对于其他情况，原型的抛弃会有利于新的最终设计的实现，这种方法就是抛弃原型。一个抛弃原型的例子就是快速原型法。在这个方法中，系统的简单版本在开发的早期就被很快搭建起来。这样的原型也许仅仅包含少量界面图片来展示系统怎样与用户交互及它将具有的功能。目的不是制作产品的有效版本，而是获得一个示范工具，以便阐明有关方面之间的沟通。例如，在分析阶段解决系统的需求问题，或在销售阶段作为向潜在客户演示的辅助，快速原型具有很大的优势。

第 4 章　数据库技术

4.1　数据库原理

课文

1．数据库的基本概念

数据库是相关数据的集合。我们所指的数据是指存在的事实。这些事实是可以被记录的并且有内在的意义。例如，想想所有你知道的人的名字、电话号码和地址，你可以用索引地址簿记录这些数据，或者用个人微机及诸如 DBASE Ⅲ 或 Lotus 1-2-3 这样的软件把它存储在软盘上。这是一个具有内在意义的相关数据集合，因而它是一个数据库。

上面的数据库的定义是相当概括的。例如，我们可以考虑所有组成这页文字的词的集合，它们是相关的数据因而是数据库。然而，作为通用的术语数据库常常是更有限制的，数据库有下面内在的性质：

- 数据库是在逻辑上紧密相关的并且有某种内在意义的数据集合；随机的数据聚合不能称为数据库。
- 数据库是由用于某种特定目的的数据设计、构造和提供的。它有一些预期的用户和这些用户感兴趣的某些预想的应用。
- 数据库表示现实世界的某一方面，有时称之为"微型世界"，微型世界的改变反映在数据库中。

换言之，数据库有某种"源"，从这里数据被引导出来；有与现实世界的事件在某种程度上的交互；有对数据库的内容非常关注的使用者。

2．数据库管理系统

这些多用户数据库是由称作数据库管理系统（DBMS）的软件管理的。正是这一点使得数据库区别于一般计算机文件。DBMS 位于物理数据库（即实际存储的数据）和该系统的用户之间，所有来自用户的对数据的存取要求——不论来自终端用户还是批处理程序，都由 DBMS 来处理。

DBMS 的一般功能是把数据库用户与机器代码分隔开。换句话说，DBMS 提供了一种将数据提高到硬件层次之上的观点，并支持像 "Get the PATIENT record for patient Smith" 这样以高级语言编写的用户请求。

DBMS 还决定每个用户能够访问的数据库信息类型和总量。比如，外科医生和医院管理者将需要不同的数据库视图。

当用户想访问数据库时，要采用 DBMS 能理解的特定数据操作语言来提出访问请求。DBMS 接收到这一请求后，对其进行句法检查。接着，DBMS 依次检查外部模式、概念模式，并将概念模式映射到内部模式。然后，对存储的数据进行必要的操作。

一般地，可能需要数据库中数据的几个逻辑表的域。每个逻辑记录值可能依次需要多个保存在实际数据库中物理记录中的数据。DBMS 必须检索需要的每个物理记录，并构造用户所需数据的逻辑视图。这样，用户就无需去了解数据库的物理布局情况。由于性能原因可能要改变其物理布局，但用户无需修改数据结构的逻辑视图。

数据库管理系统是能使用户建立和维护数据库的程序集合（见图4-1）。因此 DBMS 是一种一般用途的软件系统，它可以使得为各种应用而定义、构造和操作数据库的过程变得容易。"定义"一个数据库包括指出在数据库中存储的数据类型以及对每种数据类型的详细说明。"构造"数据库是在 DBMS 控制的某种存储介质上存储数据本身的过程。"操作"数据库包含这样的一些功能，诸如查询数据库以检索特定的数据，更新数据库以反映在微型世界中的变化以及从数据中产生报告。

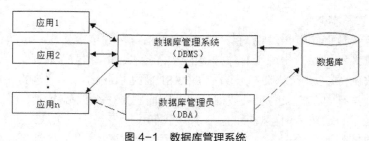

图 4-1　数据库管理系统

3．DBMS 的逻辑结构

（1）表结构

在表结构逻辑方式中，记录通过指针链接在一起。指针是记录中的一个重要数据项，它指出另一个逻辑相关的记录的存储位置。例如，顾客主文件中的记录将包含每个顾客的姓名和地址，而且该文件中的每个记录都由一个账号标识。在记账期间，顾客可在不同时间购买许多东西。公司保存一个发票文件以反映这些交易，这种情况下可使用表结构，以显示给定时间内未支付的发票。

（2）树型结构

在树型结构逻辑方式中，数据单元的多级结构类似一棵"倒立"的树，该树的树根在顶部，而树枝向下延伸。在树型结构中存在主从关系，唯一的根数据下是从属的元素或节点，而每个从属的元素或节点又依次"拥有"一个或多个其他元素（或者没有）。

（3）网状结构

网状结构不像树型结构那样不允许树枝相连，它允许节点间多个方向连接。这样，每个节点都可能有几个所有者，而它又可能拥有任意多个其他数据单元。数据管理软件允许从文件的任一记录开始提取该结构中的所需信息。

（4）关系型结构

关系型结构由许多表格组成，数据则以"关系"的形式存储在这些表中。例如，可建立一些关系表，将大学课程同任课教师及上课地点连接起来。

4．信息系统

信息系统的目标就是要在最恰当的时刻，按可接受的精确度并以经济的成本向各个管理阶层提供信息。

每个行业所需信息都视其业务特征而定。汽车制造商特别关心国内市场上来自海外制造商以及国内其他制造商竞争的程度。旅行社关心购买力及其对假日订票的影响，以及各个国家的政治局势。作为一般性的指导，信息报告的细节随接受者在管理阶层中的位置而变化。他们需要与事件发生时相关的信息以便采取适当的措施控制它们。

由于需要快速响应和灵活查询，信息系统通常要用计算机处理。在信息系统的底层是事务处理系统，它获取并处理内部信息，如销售、生产和库存数据，产生业务的工作文档如发票和财务报表。这总是公司要安装的首要系统。在事务级系统之上的是决策支持系统，它获取外部信息，如市场走向和其他外部财政数据，以及处理过的内部信息如销售趋势，产生战略方案、预测和预算。这种系统通常与个人电脑中的制表软件和其他不相连的工具放在一起。管理信息系统位居信息需求的顶层。MIS 从事务级系统获取方案和信息，从整体上控制业务运作，为战略方案、预测和预算的制定提供反馈信息，再反过来对事务级产生影响。

MIS 设计者的一个棘手的问题是研制决策支持所需的信息流。一般而言，不同级别、不同职能的管理者所需的信息大多来自现有信息系统（或子系统）集，这些系统在 MIS 中可紧密地结合在一起，但更经常的情况是松散结合。

阅读资料

典型数据库介绍

1．SQL Server

Microsoft SQL Server 是第一个可伸缩的高性能数据库管理系统，设计成满足分布式计算要求的客户机/服务器模式。Microsoft SQL Server 提供了以下功能。

- 综合了 Microsoft Windows NT 线程和进度安排服务、性能监视和事件观察功能。向 Windows NT 网络和 SQL Server 的单独登录由用户账户来简单管理。
- 建立整个企业的信息发布安全复制体系，减少了停机时的危险，并为需要信息的人及时、准确地提供信息。
- 并行体系结构。通过并行方式实现内部数据库功能，系统性能和可伸缩性显著增强。
- 用完全分布式框架对整个企业进行服务器集中管理。基于视窗的管理界面为基于多服务器的数据复制、服务器管理、诊断和调整等远程管理提供了可视化的拖放式控制。
- 通过采用并行体系结构的优越性为大型数据库提供更好的支持。为许多开发和维护任务减少了 I/O 工作。
- 分布式管理对象的 OLE 库在分布式管理框架中非常有用。

2．Oracle

作为每个数据库服务器的一部分，Oracle 包括 Oracle 企业管理器（EM）、一个带有图形接口的用于管理数据库用户、实例和提供 Oracle 环境等附加信息功能的数据库管理工具框架。

在 Oracle 9i 的 EM 版中，超级管理员可以定义在普通管理员的控制台上显示的服务，并能建立管理区域。

Oracle 企业管理器可用于管理 Oracle 标准版或企业版。在标准版中，用于诊断、调整和改变实例的附加功能由标准管理包提供。对于企业版，这些附加的功能由单独的诊断包、调整包和变化管理包提供。

正如每个数据库管理者所熟知的，对数据库做备份是一件很普通但又必要的工作。典型的备份包括完整的数据库备份、桌面空间备份、数据文件备份、控制文件备份和存档注册备份。Oracle 8i 为数据库服务器管理备份和恢复引进了恢复管理器（RMAN）。以前，Oracle 的企业备份工具（EBU）在一些平台上提供了相似的解决方案。然而，RMAN 及其存储在 Oracle 数据库中的恢复目录提供了更完整的解决方案。RMAN 可以自动定位、备份、存储并恢复数据文件、控制文件和存档记录注册。当备份到期时，Oracle 9i 的 RMAN 可以重新启动备份和恢复来实现恢复窗口的任务。Oracle 企业管理器的备份管理器为 RMAN 提供基于图形用户界面的接口。

3．网上数据库技术

网络为用户提供了访问各种多媒体数据的功能。诸如反向检索的即时信息检索技术，允许对文本进行基于关键字的高效检索，这在很大程度上满足了对网络进行访问的次数呈指数性增长的需求。随着用户方面压力的增长，即要求访问的信息类型更加丰富、要求提供不仅仅是基于简单关键字搜索的服务等，数据库研究团体给出了两种解决方案。第一种是利用数据库为网页建立模型，信息可以被提取出来用以动态地建立一个梗概，用户以此梗概为背景提交类似于 SQL 的查询。通过采用 XML 的数据表达方式，第二种方案集中在把数据库结构添加到 HTML 中去，以提供更丰富的、可查询的数据类型。

随着研究人员试图弄懂网上大量不同种类组成的、快速演化的数据的意义，今天的数据库管理系统技术还面临着另一个挑战。大量的协作式数据库使自主性与异构性问题大大复杂化了，这需要一种详细的、可扩展的方法。我们需要用更好的模型和工具来描述数据语义并规定元数据。

4.2　数据仓库和数据挖掘

课文

计算机和网络技术的进步导致出现了功能非常强大的，能收集、管理和分发大量有关数据的硬件和软件平台。在商业应用中，与产品和服务相关的交互经常会涉及繁琐的事务处理。这些事务不限于商业方面，它们也会在政府、保健、保险、制造、财经、分销、教育等方面出现。任何企业，如果有了某些计算机化的记录保存系统，并且很希望从大量详细而松散的信息池中演绎或得出逻辑结论，就应该考虑建立一个企业级数据仓库应用系统。之后，这些企业就能够提高洞察运营趋势的能力，最终提高其预测和计划的精确性。尤其是操作数据分布在多个异构系统上，并可取代人工数据收集和调节过程时，数据仓库应用系统能提高效率。

1．数据仓库

很多公司允许他们的数据被存在很多开放的系统上，但这些系统不能在公司范围内提供一个统一的可用的信息视图。解决这个问题的一个方法是建立数据仓库。数据仓库是一种数据库，它将从不同产品和操作系统调出的数据组合起来放入这种大型数据库，对管理状况做出报告和进行分析。这种数据库对源于组织核心事务处理系统的数据进行重新组织并与其他信息（包括历史数据）进行合并。这样，这些数据可以用来做出管理方面的决策和分析（见图 4-2）。

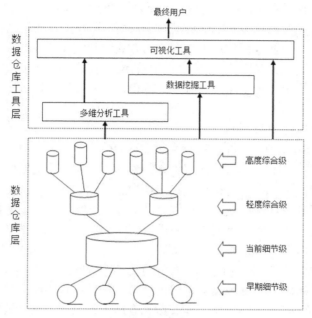

图 4-2 数据仓库系统结构

数据仓库对于不断进取的、具有竞争力的、成为关注焦点的组织来说不是一个附属品，而是不可缺少的。它是建立决策支持和执行信息系统工具的合适的基础，这些常常是用于衡量和评价某个组织是否正在向预定目标前进的标志。

操作数据是高度结构化的、支持一个组织开展日常工作以及持续发展的信息集。如果一个组织的各部门是分散的，则操作数据有时是在远程的分布式异构系统中产生的。分布式系统可以跨越不同的地域和时区。分布式系统可以配置成使商业过程具有伸缩性、可视性，并具备跟踪的能力。

支持数据仓库应用的基础设施与其他大多数应用所依靠的基础设施是相同的。不同点在于产品层次上的种类和专用性方面，这在很大程度上提高了数据仓库基础设施的质量。

下面是在数据仓库市场上最具代表性的一些技术。为了创建出最能满足用户需求的数据仓库，作为周期性资源能力计划的一部分，需要对这些重要技术进行评价。根据需要和可用的资源，可以选择和配置最佳组合。

- 服务器技术
- 客户技术
- 数据库管理系统（DBMS）技术
- 网络技术
- 海量存储技术
- 数据表示和发布技术
- 软件工程方法学和工具

在大多数情况下，数据仓库中的数据只可用来进行报告。数据不可进行更新，所以公司的隐性操作系统的表现就没有受到影响。数据仓库这种侧重解决问题的特性，使众多的公司由于运用了数据仓库而获益匪浅。

数据仓库一般有重新塑造数据的能力。关系数据库的数据视图让用户从两维观察数据。多维数据视图允许用户以多于两维的方式观察数据。例如，按地区按季度销售。为了提供这种信息，组织可以用一种特殊化的多维数据库，或用可以在关系数据库中生成数据的多维视图的工具。多维分析能够使用户使用多维的不同方式看到相同的数据。信息的每个方面——生产、定价、成本、地区或时区——都代表不同的维。所以一个产品经理能用多维工具得知七月在西南销售区共卖出了多少件，与前一个月和去年七月相比怎么样，和销售预测相比怎么样。

构造数据仓库面临的问题如下。

- 何时以及如何收集数据。在数据收集的源驱动体系结构中，数据源要么连续地在事务处理发生时传送新信息；要么阶段性地，譬如每天晚上传送新信息。在目标驱动体系结构中，数据仓库阶段性地向源发送对新数据的请求。

  除非对源的更新通过两阶段提交在数据仓库中做了复制，否则数据仓库不可能总是与源同步。两阶段提交通常因开销太大而不被采用，所以数据仓库常会保留稍微有点儿过时的数据。但这对于决策支持系统来说通常不是问题。

- 采用什么模式。各自独立构造的数据源可能具有不同的模式。事实上，它们甚至可能使用不同的数据模型。数据仓库的部分任务就是做模式集成，并且在数据存储前将数据按集成的模式转化。因此，存储在数据仓库中的数据不仅仅是源端数据的拷贝，同时它们也可被认为是源端数据的存储视图（或实体化的视图）。

- 如何传播更新。数据源中关系的更新必须被传至数据仓库。如果数据仓库中的关系与数据源中的一样，传播过程则是直截了当的。

- 汇总什么数据。由事务处理系统产生的原始数据可能太大而不能在线存储。但是，我们可以通过对关系做总计而得到的保留汇总数据回答很多查询，而不必保留整个关系。例如，我们不是存储每件服装的销售数据，而是按类存储服装的销售总额。

2. 数据挖掘

数据挖掘是关于使用自动或半自动化方法去分析数据和寻找隐蔽模式的技术。数据挖掘对企业有重要的商业价值。

- 增加竞争力
- 顾客分割
- 周转分析
- 交叉销售
- 销售预报
- 欺诈检测
- 风险管理

在过去的十年中，大量的数据积累并存储在数据库中。其中很多数据来自商业软件，如财务应用、企业资源计划（ERP）、客户关系管理（CRM）和万维网运行记录。这些数据收集的结果，使得各个组织变成数据丰富而知识贫穷者了。数据收集得非常多并且增长得很快，以至于这些存储数据的实际使用很受限制。数据挖掘的主要目的是从手头上的数据中提取模式，增加数据的自身价值，并将这些数据转变为知识。

数据挖掘对数据集提供算法，如判定树、聚类、关联、时序等算法，并分析它们的内容。这一分析产生了模式，它能给出有价值的信息（见图 4-3）。与基本算法有关，这些模式可以呈现为树形、规则、聚类或是简单的一组数学公式。在这种模式中发现的信息可用于做出报告，去指导市场策略，而更重要的是预测。

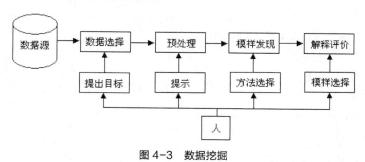

图 4-3 数据挖掘

阅读资料

专家系统

依赖人类领域专家的知识作为系统的问题求解策略是专家系统的主要特征。专家系统是使用经过编码的知识来解决专门领域中的问题的一组程序，这个专门领域通常需要人的专门知识。

专家知识必须从专家或其他专门知识源，如教科书、杂志文章和数据库中获得。这类知识通常需要在诸如医学、地质、系统配置或工程设计等一些特定领域中的许多培训和经验。一旦已收集了足够多的一批专家知识，它必须以某种形式进行编码，放进知识库，然后进行测试，并且在系统的整个生存期期间不断改进。

专家系统在以下几个重要方面与常规计算机系统不同。

1. 专家系统使用知识而不是数据来控制求解过程。大部分使用的知识实际上是启发式的，而不是算法型的。

2. 知识作为一个与控制程序分开的实体进行编码和维护。因此，它不与控制程序一起编译，这就允许增量式地增加和修改知识库，而不需重新编译控制程序。

3. 专家系统能够解释一个特定的结论是怎么得到的，以及在咨询过程中为什么需要所要求的信息。

4. 专家系统对知识使用符号表示法，并且通过符号计算进行推理，这非常类似于自然语言处理。

5. 专家系统经常用元知识进行推理，即它们用有关它们自身的知识以及它们自己的知识范围和能力来推理。

专家系统的推理应受到公开检验，提供问题求解状态的相关信息以及程序做出选择和决定的解释。对医生或工程师等人类专家来说，如果要他或她接受计算机的建议，合理的解释很重要。

人工智能和专家系统设计的探索性本质要求程序便于原型化、测试和修改。人工智能的编程语言和环境应支持这种反复迭代的开发方法。例如，在完全的产生式系统中，单一规则的改动不会影响全局的句法。规则可以添加，也可以删除，无须过多更改整个程序。专家系统的设计者们通常认为知识库是否便于修改是衡量系统是否成功的一个主要因素。

专家系统的另一特性是启发式问题求解方法的运用。专家系统的设计者发现，非正规的

"窍门"和"经验法则"是教科书和课堂正统理论的必要的补充。有时，这些规则是以可理解的方式扩充了理论知识，更多的时候仅仅是工作中经验性的捷径而已。

有趣的是大多数专家系统针对的都是相对专业性较强的、专家级的领域。这些领域通常已认真研究过，问题求解策略有明确的定义。而那些依赖于模糊定义的常识性概念的问题则很难用这些方法解决。其局限性表现在：

- 获取问题领域"深层"知识的困难性。
- 缺乏强健性和灵活性。
- 无法提供深入的解释。
- 验证的困难性。
- 不能从经历中学习知识。

尽管有这些局限性，专家系统仍在大量的应用中证明了它们的价值。

4.3 大数据和云计算

课文

1．大数据

"大数据"通常指的是所有来自不同来源的 $10^{15} \sim 10^{18}$ 的数据。换句话说，数十亿到数万亿的记录。"大数据"能够被更大量地产生，并且产生的速度远远快于传统数据。尽管每条微博被限制为 140 个字符，但 Twitter 公司每天仍然产生了超过 8 兆字节的数据。根据 IDC 技术研究公司的调研，数据每两年都会增长一倍以上，因此对组织而言可用的数据量不断激增。理解数据的高速增长以便获取市场优势至关重要。

企业对"大数据"感兴趣，是因为它们与更小的数据集相比包含了更多的模式与有意思的特殊情况，"大数据"具有能够提供新的针对顾客行为、天气模式、金融市场活动或其他现象的洞察力的潜力。然而，要从这些数据中获取商业价值，组织需要能够管理和分析非传统数据以及其传统企业数据的新的技术和工具。

数据的巨大增长与"云"的开发相一致，"云"实际上只是全球互联网的一个新的别名。随着数据的激增，特别是使用 Apple 和 Android 智能手机及操作系统的移动数据应用程序的增长，对数据传输和存储的需求也增长了。诸如 Microsoft.com 或 FedEx.com 这些网站，它们为全球的客户通信服务，有数以百万计的事务，对于从它们得到的巨大量的数据，没有单个的网站服务器配置能够提供足够的能力。用户想要快速的响应时间和即时的结果，特别是在最大和最复杂的基于网络的商业环境中，以互联网速度操作。

大数据分析需要基于云的解决方案———一种能够跟上数据传输量和速度加速增长的联网方法。这意味着服务器群和诸如 Hadoop 和企业控制语言（ECL）这些帮助找到大量数据中隐藏的含义的软件分布遍及高宽带电缆上的云处。使用基于云的实现方法，谷歌市场营销分析师可能正坐在硅谷输入信息请求，但数据是在东京、阿姆斯特丹和德克萨斯州奥斯汀同时被处理。

一些制造最复杂工业设备的大公司，现在能够从仅仅检测设备故障转到预测故障。这使得设备能在一个严重问题形成之前就被更换。

一些大的互联网商务公司，例如亚马逊和淘宝，使用大数据分析来预测买家活动以及了解仓储需求和地理定位。

政府医疗结构和医学科学家们使用大数据于早期发现和跟踪潜在的流行病。急诊室就诊突然增加，甚至某些非处方药的销售增加，可能是某种流行病的早期警告，使医生和应急部门能启动一些控制和遏制程序。

许多大数据应用程序被创建来帮助使用网络的公司与意想不到的数据量作斗争。只有最大的一些公司有大数据的开发能力和预算。但是数据存储、宽带和计算的成本持续下降，意味着大数据正迅速成为中型，甚至一些小公司的有用和负担得起的工具。大数据分析已经成为许多新的、高度专业化的商业模式的基础。

2．云技术

云计算指的是一种由"云"服务器进行的计算，这种计算通常通过互联网来实现。这种类型的网络已经投入使用多年，曾经主要用于建立技术研究和大功耗应用所需的超级计算机。在当时的情况下，它通常指的是网格计算。现在，云计算指的是通过个人计算机、手机以及其他可以登录互联网的设备来访问基于 Web 的应用程序和数据。云计算的概念为：在任何地方、任何设备上都可以使用的应用和数据。例如，当在线存储或访问文档、照片、视频和其他媒体时，当在线使用程序或应用（如电子邮件、办公软件、游戏等）时，当在线与他人分享观点、想法以及内容（如社交网站）时，你都会用到云计算的功能。

采用云计算这个名词的灵感来自于在流程图和示意图中经常用这个云符号表示 Internet。云可分类为公共云、私有云和混合云。企业通常会使用公有云上可用的应用；它们也经常会创建私有云来管理自己公司的数据和应用。

云计算依赖于资源共享以达到聚合性和规模经济。这类似于网络上的公用设施（如电网）云计算的基础是汇聚的基础设施和共享服务这个更广义的概念。

云计算也专注于使共享资源的效用最大化。云资源通常不仅仅为多个用户共享，而且也按需被动态地重分配。它能够把资源分配给用户。例如，一个在欧洲工作时间用特定应用程序（如 email）服务欧洲用户的云计算机设备可以在北美工作时间用不同的应用程序（如 Web 服务器）重新分配该相同的资源服务北美用户。这种方法将使计算能力的使用最大化，从而也减少对环境的破坏，因为对于各种各样的功能只需要较少的电力、空调、架子的空间等。用云计算，多个用户能够访问单个服务器来检索和更新他们的数据，而不需要为不同应用购买多个许可证。

云计算提供者按三种基本模式提供他们的服务：基础设施作为服务（IaaS），平台作为服务（PaaS）和软件作为服务（SaaS）。

基础设施作为服务，如亚马逊 Web 服务用唯一的 IP 地址和按需分配的一些存储块提供虚拟服务器实例。客户使用提供者的应用程序接口（API）来开始、停止、访问和配置他们的虚拟服务器和存储。在企业里，云计算允许一个公司仅仅支付所需容量的费用，并且一旦需要可在线获得更多的容量。因为这种"支付你所用的"的模式很像消费电、燃料和水的方式；它有时称为公共设施计算。

在云中，平台作为服务被定义为放在提供者的基础设施上的一套软件和产品开发工具。开发人员通过 Internet 在提供者的平台上创建应用程序。PaaS 提供者可以使用一些 API，Web 门户网站或安装在客户计算机上的网关软件。一些 Google App 是 PaaS 的例子。某些提供者不允许他们的客户创建的软件搬离提供者的平台。

在软件作为服务的云模式中，供应商提供硬件基础设施、软件产品，并且通过前端门户与用户交互。SaaS 是一个很广阔的市场。服务可以是任何一种，从基于 Web 的 email 到库存控制和数据库处理。因为服务提供者既拥有应用程序，又拥有数据，因此终端用户可以自由地从任何地方使用该服务。

阅读资料

计算机科学的前沿技术

纳米技术

纳米技术涉及创建纳米级的器材和物质，因此，微电子学将自然演进到纳米电子学。正如微电子学导致了微型芯片或集成电路的产生，可以令人信服的预测，纳米电子学将会导致纳米芯片取代微型芯片。2007 年 6 月 26 日，IBM 宣告其第二代的最新超级计算机问世。蓝色基因是世界上运行得最快的计算机，其速度达到每秒 360 万亿次浮点运算。但是其用到的 Power 5.6 型芯片还不能算"纳米芯片"。

基于组件开发

基于组件开发（CBD）将使软件产业发生革命。与诸如对象和客户/服务器等最新趋势不同，CBD 不只是一种分布计算的新花样，而是一种可扩展的体系结构，支持整个生命周期计算的理念，包括设计、开发和部署。由于 CBD 具有高度的可重用性和互操作性，所以它将影响应用程序构成的各个方面，包括所有类型的客户机、应用程序服务器和数据库服务器，将对应用程序开发的各个方面产生深刻的影响。

CBD 的前身是面向对象的开发。CBD 为了编制和装配应用程序，按基于互操作框架的标准化基础结构的上下行文，重新定义了对象，也定义了由组件架构和框架所预定的预制组件。由于 CBD 基础设施能够实现组件的设计、建造与装配（不只是提供一种直观编程环境），它加上其可获得性将永远改变应用程序的开发方法。CBD 可以分为三个阶段：单片式、分布式和持续式。随着时间的推移，某些标准将变得非常深入普遍，核心服务将商品化，CBD 基础架构方面将保证满足可移植性和可互操作性的要求。

智能机器人

大部分现存的机器人离具有智能还相差甚远，它们不能听、看和感觉。现存的机器人不能适应其工作计划和工作环境的改变。它不会自动地改变其行动计划来适应变化的环境，只能等待人类的干涉否则就陷入混乱。设想这种情况就像一个危险而又鲁莽的家伙面对着一种复杂而又快速变化的环境那样。这些情况都告诉我们为什么需要智能机器人。

智能机器人，基本上来说就应该是一种这样的机器人，它能够感知周围的环境并具备足够的智能去对变化的环境做出响应，就像我们所做的那样。这种能力需要直接使用感官知觉和人工智能。

对机器人技术已经进行过很多研究，并且仍然很关切为机器人配备视觉传感器——"眼睛"和触觉传感器——"手指"。人工智能将使机器人对赋予它的任务和所在的环境变化做出反应和进行适应，并对这些变化进行推理和决策。在这个意义上，智能机器人可以被描述成一个这样的机器人：它除一般的机动能力外，还配备有各种传感器、具备有复杂的信号处理能力和具有做决策的智能。

普适计算

普适计算带给我们更容易地管理信息的工具。信息是全球经济的新货币。我们越来越依赖于个人的、财务的以及其他保密信息的电子化创建、存储和传递，并且所有这些交易都要求达到最高的安全级别。我们要求能完整地访问时间敏感性数据，而不管其物理位置。我们期望所有的设备——个人数字助手、移动电话、办公室 PC 和家庭娱乐系统——都能访问到那些信息，并且能在无缝连接的集成化的系统环境下一同运作。

普适计算的目的是使人们能够使用新型的智能化的和便携式的设备完成越来越多的个人

和专业的事务。它带给人们的是便利地获取存储在强大的网络上的相关信息,从而允许人们在任何地点、任何时间很容易地采取行动。

第 5 章　计算机网络技术

5.1　计算机网络基础

课文

计算机网络的建立是为了满足人们以即时方式共享数据的需求。个人计算机在处理数据、电子表格、图形以及其他类型的信息方面是理想的办公设备,但却不支持快速(用户输出的)数据共享。没有网络,就必须将文档打印出来才能供他人编辑或使用。顶多是使用软盘将文件送达其他人员,并复制到他们的计算机上。如果对方做了修改,原文件将无法反映出这些变化。这种工作模式一直被称为单机环境。

如果用户能把他的计算机与其他计算机连在一起,他就可以共享其他计算机上的数据和设备,包括高性能的打印机。一组计算机和其他的设备连接在一起构成的系统叫作网络,互连的计算机共享资源这样一种技术概念叫作网络技术(见图 5-1)。

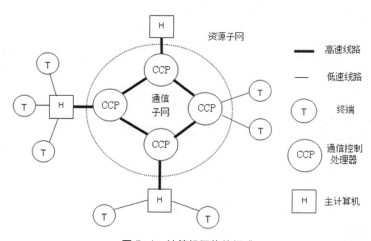

图 5-1　计算机网络的组成

网络中的计算机可以共享数据、消息、图形、打印机、传真机、调制解调器、光盘、硬盘以及其他数据存储设备。

1．网络结构

网络以各种大小和复杂性出现。总的来说,网络越大,就越复杂,也越难于管理。

(1)局域网

局域网是一种同质的实体。放在网络上的每个数据包对每个节点(服务器或工作站)都是可见的,但每个节点只处理那些发送给它的数据包,除了网络测试设备和黑客工具,它们会阅读任何自己感兴趣的数据包。局域网上的所有节点接入到一根电缆线(旧式的同轴电缆以太网)或中心集线器上。

(2)桥式局域网络

桥式局域网络是一种因为过于庞大而必须使用"网桥"来削减通信量的局域网。网桥知道它的每一边的网络地址,并仅仅将标定地址的信息送到网桥的其他地方,这样就"分割"

了整个网络。

(3) 交换式局域网

交换式局域网有一个称之为"交换机"或"交换集线器"的快速、多端口桥。交换式局域网是极其细化的，所以每个端口仅有唯一节点，这样就尽可能地降低了每根电缆上的信息量。

(4) 校园网

校园网至少有一个主干，并有两个或更多的局域网通过网桥、交换机或路由器连接到主干网上。

(5) 路由网

路由网是单独的局域网通过路由器连接起来的一种网络，路由器是比网桥或交换机更智能化的设备。路由器知道信息的位置，虽然要经过几个路由器的距离。它将信息保存到路由表中，也知道几条前往相同目的地的路由线路，并且聪明到可以算出来哪条路由线路最快。

(6) 广域网

广域网是一种可以通过非网络拥有者的链接的一种网络。总的来说，这些链接由电话公司或其他一些公共运营商持有。互联网是世界上最大的广域网，由数以千计的局域网和数以百万计的路由器和服务器组成（见图 5-2）。

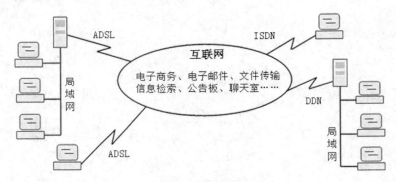

图 5-2　互联网

2．网络管理员

网络管理员对一个网络上的所有资源拥有完全占有权和许可权。网络管理员通常负责安装、管理和控制服务器和网络组件。网络管理员还可以修改用户账号的属性、工作组的成员资格，创建、管理和安装打印机、资源共享以及为那些资源分配权限。数据库管理员负责在一个网络环境中编程和维护一个大型的多关系型数据库，并且为个人在网上直接访问数据库提供便利。工作组管理员负责解决问题、实现标准和解决方案、检查性能以及提高连接到一个更大的网络环境的特定的一组人员的效率。技术支持人员负责为在大型复杂的网络环境中工作的系统管理员提供技术帮助，为终端用户提供常规问题解答和现场培训。维修承包商负责对硬件的维修和升级。Web 站点管理员负责实现和维护公司 Internet 站点的内容和风格，保证网站信息正确、及时和有趣。

3．拓扑结构

物理拓扑结构描述的是网络的布局，就像一张地图展示出各种道路的布局。而逻辑拓扑结构则描述数据是如何通过网络发送的，就如同汽车在地图上的每条道路是如何行驶的（方向和速度）。逻辑拓扑结构是一种在计算机之间传递信息的方法。

（1）星型网络

星型拓扑结构的计算机网络最简单，由一个中心计算机或集线器构成。中心计算机或集线器作为路由器通过存储转发或交换系统在被连接的计算机之间传输信息。星型拓扑结构的节点中包括集线器节点，该节点又有自己的下一级节点，在这种情况下，网络中的两个网络节点间可能存在多层路由线路。

（2）网状网络

网状网络是一种计算机网络，它以网状拓扑结构将许多计算机系统连接在一起。在规则的网状拓扑结构中，网络中的每个节点用一维或多维的两个相邻节点连接起来。如果网络是一维的，那么节点链就会被连接形成一个环路，也就是熟知的环形网络。诸如 FDDI 光纤分布式数据接口这样的网络系统就使用反向双令牌环网来获得高可靠性和性能。总的来说，当 n 维网状网络的电路连接超过一维，最后所得到的网络拓扑结构将是一个环轮，该网络也称为环轮形的。

（3）总线型网络

通过总线拓扑结构，所有的工作站都直接连接到传输数据的主干线上。任何一台计算机所产生的通话将会通过主干线传输，并且被所有的工作站接收。这种方式在由 2～5 台计算机构成的一个小型网络中能够很好地工作，但是随着计算机数目和网络通信量的增加，这种方式将会降低网络性能和可用的带宽。

在总线型网络方案中，工作站沿一根电缆（通常是同轴电缆）像珠子一样串成一串。网络的任何地方出问题则整个网络都会瘫痪。由于网络中有那么多的连接器和可能的故障点，因此故障诊断是不可靠和困难的。

（4）环型网络

环型网络的节点按闭合回路或者圆环方式连接。环形网络中的信息沿一个方向传递，从一个节点传至另一个节点。因为信息绕环传递，故每个节点都可检查信息所属的终点地址。如果地址与节点所配地址一样，则节点接受该信息；否则，它会重新生成该信息，并且将信息继续沿着环传至下一个节点。实际上，通常会有两条电缆，这样适配器就会在网络中出现故障时，可以沿环路返回重新构成环。在实现上，大多数的环都是虚拟的，真正的布线就类似于星型网络模式。

（5）混合型网络

对于混合型拓扑结构，两种或多种拓扑结构结合形成一个完整的网络。例如，一个混合型拓扑结构可以是星型拓扑结构和总线型拓扑结构的结合。这也是最常用的方式。

在一个星型-总线型拓扑结构中，几个星型拓扑结构网络连接到一个总线接口上。在这种拓扑结构中，如果一台计算机出现故障，它不会影响网络上的其他设备。然而，如果在星型结构中连接所有计算机的中央设备或集线器出现故障，那么就会有大麻烦了，因为没有计算机可以互相通信了。

在一个星型-环型拓扑结构中，计算机如同在一个星型网络中那样被连接到一台中央设备上，而这些设备则被连接形成一个环型网络。与星型-总线型拓扑结构一样，如果一台计算机出现故障，将不会影响到网络上其他设备。通过使用令牌传递，在一个星型—环型拓扑结构中的每台计算机都有同等的通信机会，这使得在各段之间进行的网络通信的通信量比在星型—总线型拓扑结构中要大。

阅读资料

光纤通信

通信系统从一个地方向另一个地方传送信息，两地之间有几公里或越洋之遥。信息通常由电磁载波携带，载波的频率从几 MHz 到几百 THz。光通信系统的载波为电磁波可见光或近红外光的高载波频率。有时候称光纤通信系统为光波系统，以区别于微波系统，微波通信系统的载波比光纤通信系统的载波小 5 个量级（1GHz）。光纤通信系统是光波系统，利用光纤完成信息传输。自 1980 年以来，有些系统已在全世界有效使用，它是对电信技术的一次彻底的革命。

微波无线电用于广播和电信传输，相对于低频来讲，它们的波长更短，方向性强的天线尺寸更小，从而更容易实现。微波波段比其他射频频段有更大的带宽，可利用的频率从低于 300MHz 到高于 300GHz。特别地，微波用于电视新闻时，可以利用特殊装备的通信车将电视信号从遥远的地方传输到电视台。

光纤通信系统与微波系统的区别原则上只有载波频率范围的不同。光载波频率的典型值是 200THz，相比之下，微波载波的频率为 1GHz。仅仅因为用于光波通信的载波频率高，人们就预计光通信系统信息容量将增长 10000 倍。因为已调载波的带宽只是载波频率的百分之几，所以这个容量增长值也是可以理解的。如果将载波带宽的最小值设定为载波频率的 1%，光通信系统的比特率有可能达到 1Tbit/s。光通信系统的潜在带宽是巨大的，它是世界范围内开发和实施光波系统的驱动力。现有的最佳系统的比特率是 10Gbit/s，这表明还有很大的提升空间。

光纤通信基本上适用于任何需要将信息从一个地方传送到另一个地方的领域。然而，光纤通信系统在电信方面的应用最成熟。鉴于遍布全世界的电话网就不难理解，电话网除了传输语音信号外，还传输计算机数据和传真信息。电信应用可大致分为两类：长途和短途（本地），这个分类取决于将光信号传输的距离与城市间的距离（100km）相比是长还是短。长途通信系统要求大容量中继线和获益颇丰的光纤通信系统。确实，光纤通信背后的技术发展常常受长途通信的驱动。

每一代光波通信的成功都是在更长的传输距离上实现了更高的比特率。利用中继器将光信号周期性地放大仍需要很多长途传输系统。然而，与同轴电缆系统相比，光波系统的中继距离和比特率增长了一个数量级，因此光波系统在长途通信中有更强的吸引力。此外，传输距离为几千公里时可以利用光放大器来实现。大量越洋光波系统已经用于国际光纤网络。

短途通信涵盖市内和本地通信。这类系统主要是在 10km 距离内低比特率传输信号。使用单信道的光波系统费用较低，也应该考虑多业务的多信道网络。宽带集成业务数字网络的概念要求高容量系统，能实现多种业务。ATM（异步传输模式）技术也要求较高的宽带。只有光纤通信系统可能满足这种宽带分布的要求。

5.2 信息安全

课文

对一个网络的安全威胁可分为那些含有某种形式的未经授权访问和所有的其他情况。一旦有人得到了对网络的未经授权的访问，那么他们可以做事的范围就很大了。一些人只是对突破安全的挑战很感兴趣，而没有采取进一步行动的兴趣。他们可能选择监控网络流量，即所谓的窃听，其目的是为了了解一些特别的事情或者将其泄露给其他人，或者为了分析通信模式，通信量分析可以得到观测数据。我们能经常想起的这类未授权访问都是主动安全攻击，在一些人得到了对网络的未授权访问后，他们就采取一些明显动作了。主动攻击包括修改消息内容、冒充他人、拒绝服务和种植病毒。

管理者的责任

网络安全政策是管理部门对网络安全的重要性及他们所承担义务的声明。该政策需要用一般条款来描述应该怎样做，但并不涉及到保护目的的方法。编写这个政策是非常复杂的，因为在现实中，它通常是一个更宽泛的文件，即该组织的信息安全政策的组成部分。网络安全政策需要明确申明管理层对网络安全和需要保护内容非常重要的立场。管理层必须明白一个绝对安全的网络是不可能存在的。

此外，由于科技的进步和那些想入侵网络和计算机的人的创造性，网络安全是一个不断变化的目标。今天，为最大限度地减少安全风险而落实的措施在将来还需要升级，通常升级需要附加资金，管理者需要为之支付。

简单写写政策并不能将实践、过程或软件顺利实施来改善安全状况。这需要后续通过与所有员工的沟通，使他们认识到高层管理人员在安全上布置的重点和重要性。在各个层面，管理层都需要去支持这项政策，并定期与员工用各种方式强化它。IT和网络的工作人员可能需要安装额外的硬件、软件和程序来自动进行安全检查工作。

攻击预防

不同的安全机制可被用来加强既定安全策略限定的安全性能。根据预期的攻击，要使用不同的手段来实现想要的性能。攻击预防是安全机制的一个类别，它包括在攻击真正到来并影响目标之前进行预防或阻止的一些方法。这类方法的一个要素就是控制访问，即可以应用于如操作系统、网络或应用层等不同层次的一种机制。

访问控制限定并控制对重要资源的访问。它是通过识别或验证资源请求方并核准其对特定对象的权限来实现的。攻击者没有合法的许可去使用目标对象因而被拒绝访问资源。由于访问是攻击的前提，所以任何可能的干扰得以预防。

在多用户计算机系统中，最常用的访问控制形式是资源访问控制清单。清单是建立在进程中想要使用这些资源的用户和群的身份基础上的。用户身份由最初的验证过程决定，通常要一个用户名和密码。登录过程检索同用户名对应的存储的密码备份并同递交的密码进行比较。当两者都匹配时，系统给予相应的用户和群证书。当一个资源被请求访问时，系统会在访问控制清单中查找用户和群并适当授权访问或拒绝访问。在UNIX文件系统中可以找到这种访问控制的例子：文件系统基于用户和群的身份提供可读、可写、可执行的许可。在这个例子中，对于没有被授权使用文件的用户对文件的攻击就可以因操作系统中文件系统代码的访问控制部分而得以预防。

防火墙是网络层中一个重要的访问控制系统。防火墙的思想是基于将独立管理控制下的可信任的计算机内部网络同有潜在威胁的外部网络分开。防火墙是一个主要的阻塞点，能加强对网内或网外服务的访问控制。防火墙通过拒绝来自外部的未授权方的连接企图，从而防止来自外部的对内部网络中的机器的攻击。除此之外，防火墙还可以用来阻止墙内用户使用外部服务。

攻击避免

这类安全机制认为入侵者可以访问到想要的资源，但信息可以以某种方式修改使得攻击者无法使用。信息在通信渠道中传播之前由发送者进行了预处理，并且接收者收到信息后也要进行处理。当信息在通信渠道传送时，对入侵者来说几乎是无用信息，从而抵制了攻击。一个值得注意的例外是对信息的可用性的攻击，因为攻击者仍然可以中断信息。在接收者处理信息的阶段，已经发生的更改或错误是可以被检查出来的（通常是因为信息不能正确重构）。当没有发生更改时，接收者处理后的信息同发送者进行预处理前的信息是一样的。

这种方法中最重要的成员是密码学，一门保持信息安全的学科。它使得发送者将信息进行转化，转化了的信息在攻击者看来是一种随机的数据流，但是可以很容易的被授权了的接收者解码。

原始信息叫作明文。将信息通过一些转换规则的应用转换成隐藏实质的格式的过程叫作加密。相应的被隐藏了的信息叫作密文，密文转换成明文的过程叫解密。值得注意的是，明文转换成密文必须是无损的，以便在任何情况下接收者都能恢复原始信息。

转换规则通过密码算法来描述。这种算法的作用基于两个主要的原则：替换和移位。在替换中，明文的每一个要素（如字节、块）被映射成所用字母的另一要素。移位是指明文要素重组的过程。大部分系统有多个步骤的替换和移位以便使得系统有更强的抵抗解密的能力。密码解析就是破解密码的学科，即发现信息隐藏下的实质。

最常见的攻击叫作已知明文攻击，是由同时获取密文和相应的明文引起的。加密算法必须足够复杂，这样即使密码破解者同时获得大量密文和明文也无法获取密钥。当破解密码的价值超出了信息价值时，或者破解所花时间超出了信息本身的使用期限时，攻击就不可行。如果同时有多对密码和明文，很显然一个简单的猜密钥算法在一段时间后总能成功。因为没有使用任何关于算法的信息，这种不断尝试不同密钥值直到找到正确密钥的方法被称作暴力攻击法。有效的加密算法的一个必要条件就是使暴力攻击对其不可行。

阅读资料

互联网安全措施

（1）网络防火墙

网络防火墙的目的相当于在网络周围设置一层外壳，用于防止连入网络的系统受到各种威胁。防火墙可以防止的威胁类型包括：

- 对网络资源的非授权访问——入侵者可能侵入到网络的主机，并对文件进行非授权访问。
- 拒绝服务——例如，网络外的某人可能向该网络上的主机发送成千上万个邮件消息，企图填满可用的磁盘空间，或者使网络链路满负荷。
- 伪造——某人发出的电子邮件可能已被别有用心的人篡改，结果使原发件人感到难堪，或受到伤害。

防火墙可以通过滤掉某些不安全的网络服务而降低网络系统的风险（见图5-3）。例如，网络文件系统（NFS）服务可以通过封锁进出网络的所有NFS流通而防止为网络外部人员所利用。这就保护了各个主机，使其仍可在内部网络上提供服务，这在局域网环境中很有用。一种避免与网络处理相关问题的方法是把组织机构的内部网与其他外部系统完全断开。当然这不是一个好办法。所需要的办法是对访问网络时进行过滤，同时仍允许用户访问"外部世界"。

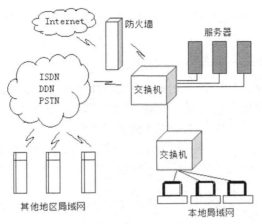

图 5-3　网络结构实例

在这种配置中，用一个防火墙网关把内部和外部网分开。网关一般用于实现两个网络之间的中继服务。防火墙网关还提供过滤服务，它可以限制进出内部网络主机的信息类型。有三种基本防火墙技术：包过滤、电路网关和应用网关。通常可采用上述的多种技术以提供完整的防火墙服务。

（2）**数字证书和认证**

数字证书相当于电脑世界的身份证。当一个人想获得数字证书时，他生成自己的一对密钥，把公钥和其他的鉴定证据送达证书授权机构（CA），证书授权机构将核实这个人的证明，来确定申请人的身份。如果申请人确如自己所声称的，证书授权机构将授予带有申请人姓名、电子邮件地址和申请人公钥的数字证书，并且该数字证书由证书授权机构用其私有密钥做了数字签名。当 A 要给 B 发送消息时，A 必须得到 B 的数字证书，而非 B 的公钥。A 首先核实带有证书授权机构公钥的签名，以确定是否为可信赖的证书。然后，A 从证书上获得 B 的公钥，并利用公钥将消息加密后送给 B。

（3）**数字签名**

多少年以来，文件上的特殊封条或手写签名当作是作者身份的证明或者对文件内容的认可。有些属性使得手写签名的使用不能被替代。这主要包括如下内容。

● 一个签名是不能伪造的，它作为签名者有意在文件上签名的证明。
● 一个签名是可信的，它使得接收者相信签名者有意在文件上签名。
● 一个签名不是可重用的。它是文件的一部分，一个不道德的人不能将它转移到另一个不同的文件上。
● 一旦签了名，该文件就不能更改。
● 一个签名不能被否认。因为签名和文件是实际的对象，之后签名者不能声称他或她没有签名。

一个数字签名就是一种电子签名，它被用来鉴别信息发送者或文件签名者的身份，用于确保已发送的信息或文件的原始内容没有被改动。一个数字签名是被附在一份电子文档上的一个二进制位串，该文档可能是一份文字处理文件或者是一封电子邮件。这个二进制位串由签名者生成，它是基于文件的数据和个人的保密口令。数字签名易于传输，不能被其他人所假冒，而且可以自动地被打上时间戳。接收文件的人能证明签名人确实在文件上签名。如果文件被改动，签名者还可以证明他没有在被改动的文件上签名。一个数字签名可以用于任何类型的信息，不论该信息是否加密，因此接收者能确信发送者的身份以及收到的信息完好无缺。

5.3 无线网络

课文

无线技术，简单来说，就是使一个或多个没有物理连接的设备进行通信，即不需要网络或外围电缆。无线技术使用无线电频率传输作为传送数据的手段，而有线技术使用电缆。无线电技术的范围从复杂的系统如无线局域网和手机到简单的设备如无线耳机、麦克风以及其他不用来处理或储存信息的设备。还包括像遥控器、无线计算机键盘和鼠标、无线高保真立体声耳机这样的红外线（IR）设备，所有这些设备需要一个介于发射机和接收机之间的直的瞄准线来关闭连接。

无线网络充当设备间以及设备与传统的有线网络（企业网和互联网）之间的传送机构。无线网络多种多样，但通常根据它们的覆盖范围分为三类：无线广域网（WWAN）、WLANs 和无线个人局域网（WPAN）。WWAN 包括宽覆盖范围技术，例如 2G 蜂窝、蜂窝数字数据包（CDPD）、全球移动通信系统（GSM）和移动技术。WLAN 包括 802.11、超级局域网等技术。WPAN 代表无线个人局域网络技术，例如蓝牙和红外线。所有这些技术都是无线的——它们利用电磁波来接收和传递信息。无线技术使用的波长范围从无线电频率波段到超过红外线波段。无线电频率（RF）波段的频率覆盖了电磁辐射频谱的重要部分，从最低的分配无线通信频率的 9 千赫兹延伸到数千吉赫兹。随着频率增加到超过了无线电频率频谱，电磁能进入到红外线中，然后进入可见光谱。

无线局域网比传统的有线局域网有更大的灵活性和可移植性。与需要一根线把用户的计算机与网络连接的传统局域网不同，无线局域网使用一个接入点装置把计算机和其他组成部分连接到网络。接入点与装在无线网络适配器上的装置通信；它通过一个 RJ-45 的端口连接到一个有线的以太局域网。接入点装置通常可达到 300 英尺（大约 100 米）的覆盖范围。这个覆盖范围被称作一个单元或域。用户可以在该域内自由地移动他们的笔记本或其他网络设备。接入点单元可以连接在一起使得用户可以在一座大楼内或大楼之间"漫游"。

特殊网络比如蓝牙被设计成能动态地连接远程设备如手机、笔记本和掌上电脑的网络。这些网络之所以"特殊"是因为它们易变的网络拓扑结构。无线局域网使用一个固定网络基础结构，特殊网络依靠一个通过无线电线路连接起来的主从式系统实现设备间的通讯，从而维持随机网络结构。在蓝牙网络中，微型网（piconet）的主体控制这些网络的不断变化的网络拓扑结构。它还控制能够支持相互直接连接的设备之间的数据流。由于设备以一种不可预知的方式移动，这些网络必须被迅速重组来处理动态拓扑。协议蓝牙使用的路由选择允许主体建立并保持这些易变网络。

近年来，无线网络变得越来越有用、负担得起并易于使用。家庭用户大量地使用无线技术。忙碌的便携式电脑用户们经常能在像咖啡店或机场之类的地方找到免费的无线连接。

无线网有许多用途。例如，在旅途中可使用计算机发送邮件、接收电话和传真以及阅读远程文件。另外，无线网对卡车、出租车、汽车和维修人员与基地保持联系极其有用。尽管无线网容易安装，但也有一些缺点。典型地，无线网容量较低，它比有线局域网慢得多，差错率通常也比较高，并且不同计算机的传输可能会互相影响。

如果你正在使用无线技术，或考虑开始使用无线技术，你就应该了解你可能会遇到的安全威胁。如果你不能确保你的无线网络的安全，任何拥有一台能激活无线网络的计算机的人只要在你的无线接口范围之内就能通过你的无线连接免费上互联网。通常室内广播接入点范围在150~300英尺。在户外，这个范围也许会扩大到1000英尺那么远。所以，如果你的邻居住得很近，或是你住在一间公寓或单元住宅里，如果无法确保你的无线网络安全就会使你的互联网连接向数量惊人的用户开放。这样做会引发一系列问题。

- 服务侵害：可能会超过你的互联网服务提供商所允许的连接数量。
- 带宽不足：背负在你的互联网连接上的用户可能会耗尽你的带宽并使你的连接缓慢。
- 恶意用户滥用：背负在你的互联网连接上的用户可能会从事非法活动而追溯到你的头上。
- 监视你的活动：恶意用户能够监视你的互联网活动并盗取密码和其他敏感信息。
- 直接攻击你的计算机：恶意用户能够访问你计算机中的文件，安装间谍软件以及其他恶意程序，或控制你的计算机。

驾驶攻击是一种特殊的背负式连接。无线接入点的广播范围使你家以外的连接或离你街道很远的连接成为可能。精明的计算机用户了解这一点，一些人已经形成了一个业余爱好，他们带着一个装有无线设备的计算机驾车穿过城市和邻近地区时——有时带一个有威力的天线——搜索不安全的无线网络。这种行为被人们戏称为"驾驶攻击"。驾驶攻击者们常常记下不安全的无线网络的位置并在网站上公布这个信息。心怀不轨的人进行驾驶攻击是为了找到一个连接，用它来从事一些非法在线活动并掩盖自己的身份。他们也会直接攻击你的计算机。

无线接入点能对无线激活的计算机敞开大门。这被称为"标识符广播"。在某些情况下，标识符广播是可取的。比如说，一家互联网咖啡馆希望它的顾客很容易地找到它的接入点，因此它会使它的标识符广播激活。当你使用一个无线路由器或接入点去建立一个家庭网络时，你用有线连接交换到由无线电信号传送的连接。除非你可以保护这个信号的安全，否则陌生人能搭上你的互联网连接，或者更糟糕的是，通过你硬盘上的文件来监控你的在线活动。

许多公共接入点并不可靠，而且它们传送的通信量也没有经过加密。这能使你的敏感通信或交易处于危险境地。由于你的连接被畅通无阻地传送出去，恶意用户能够使用"嗅探"工具来获得敏感信息，诸如密码、银行账号以及信用卡号。

正如同不安全的本地无线网络一样，一个不安全的公共无线网络与不安全的文件共享结合会招致灾难。在这些情况之下，一个恶意用户能进入到你允许共享的任意目录和文件。通过一个公共无线接入口访问互联网时，你还需要防范一些严重的安全威胁。由于你无法控制无线网络的安全设置，这些威胁是多重的。而且，你经常身处于无数能激活无线网络的计算机的范围之中，而这些计算机的主人却不得而知。

阅读资料

互联网应用

- 电子邮件

电子邮件是指利用通信网络传送信息。电子邮件类似于办公备忘录，一则信息包括信头和正文两部分。信头用于说明发信人、收信人及主题等，而正文则是信件的文本。要使用电子邮件，用户必须申请一个信箱，也就是一块存储空间，用以存放邮件。

电子邮件地址实际上是由@字符（读作 at）分成两部分的一个字符串。第一部分称作信箱标识符，第二部分则说明了信箱所在的计算机域名。信箱标识符是由本地分配的，只有在指定计算机上才有意义。有些计算机系统规定用户的信箱名与其注册账号一致；在另外一些系统上，信箱名与账号则是无关的。邮件地址中的计算机名实际上就是计算机的域名。

● 搜索工具和方法

搜索工具就是执行搜索的计算机程序。一种搜索方法就是一个搜索工具从它的 Web 网站请求和检索信息的方法。一个搜索是从所选择的搜索工具的网站开始的，通过它的地址或 URL 进行搜索。基本上有 4 种搜索工具，每一种都有它自己的搜索方法。

（1）目录搜索工具通过主题搜索信息。

（2）搜索引擎工具通过使用关键字搜索信息并得到一张参考表。

（3）带搜索引擎的目录，同时使用上面所述的主题和关键字搜索方法。

（4）多引擎搜索工具并行使用多个搜索引擎。

每一种搜索引擎有它自己适合的赋值方法。检索一个网站的数据库有以下 3 种方法。

（1）全文检索。

（2）关键字检索。

（3）个人检索。

● 移动互联网

移动互联网指的是使用诸如智能手机、功能手机等手持移动设备，通过连接移动网络或其他无线网络来接入万维网，即基于浏览器的互联网服务。

从传统上来说，接入万维网一直是通过固定线路服务，并将结果显示在笔记本电脑和台式机的大屏幕上。然而，万维网正在变得更加易于通过便携和无线设备接入。2010 年国际电信联盟（ITU）的报告说，以当前的增长速度，通过笔记本电脑和智能移动设备接入万维网的移动中的用户的数量，将会在未来的 5 年内超过使用台式机接入的人数。随着自 2007 年以来更大的多点触控智能手机的兴起，以及自 2010 年以来的多点触控平板电脑的兴起，向移动互联网方向的转变正在加速。这两种平台都提供了比上一代移动设备更好的互联网接入、屏幕、移动浏览器，或基于应用程序的用户 Web 体验。Web 设计者可能会单独制作这样的页面，或者页面也许可以像在移动维基百科中一样被自动转化。

可以预见，移动 Web 应用（移动应用）和本地应用的区别变得越来越模糊，因为移动浏览器可以直接接入移动设备（包括加速计和 GPS 芯片）的硬件。因此，基于浏览器的应用的速度和能力得以提升。针对高级的用户界面图形功能的持续存储和访问会进一步减少对开发特定平台的本地应用的需求。

至今，移动互联网接入仍然受到互用性和可用性问题的困扰。互用性问题源于移动设备、移动操作系统和浏览器的平台的多样性。可用性问题围绕着移动电话规格小的物理尺寸（对显示器分辨率和输入/操作的限制）。尽管存在这些缺点，很多移动开发方还是选择使用移动互联网来创建应用。

移动互联网指的是通过手机服务提供商来接入互联网。这种无线接入可以在手机切换服务区的时候，切换到别的无线电发射塔。它可以指持续连接同一个信号塔的固定设备，但是，在这里它不是"移动"的意思。通常来讲，静止用户可以使用 Wi-Fi 和其他更好的方式。相比之下，通过电话系统的蜂窝基站比直接连接互联网服务提供商的无线基站所提供的服务更加昂贵。

诸如智能手机等不通过蜂窝基站的方式，能够连接数据或语音服务的移动电话，不在移动互联网上。而配有宽带调制解调器和蜂窝服务提供商注册的笔记本电脑（正穿越城市在公交车上旅游），则在移动互联网上。

移动宽带调制解调器可将智能手机"绑定"在一个或多个计算机，或其他终端用户的设备上，通过那些蜂窝电话服务提供商所能提供的协议，从而提供对互联网的接入。

第6章 电子商务

6.1 电子商务

课文

电子商务就是利用电子媒介做生意（见图 6-1）。它意味着利用简单、快速和低成本的电子通信实现交易，无需交易双方见面。现在，电子商务主要通过互联网和电子数据交换（EDI）的方式来实现。电子商务于 20 世纪 60 年代提出，随着计算机的广泛应用，互联网的日趋成熟和广泛利用，信用卡的普及，政府对安全交易协议的支持和促进，电子商务的发展日臻繁荣，人们开始利用电子媒介做生意。

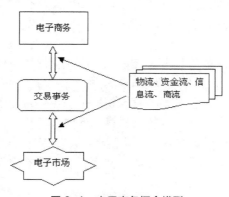

图 6-1 电子商务概念模型

计算机网络提供了快速廉价的信息交换方式，互联网几乎进入了世界的每一个角落。中小企业可与世界任何地方的合作伙伴建立全球性的关系。高速的计算机网络使地理上的距离变得微不足道。商家可以将商品卖给非传统市场的客户，开辟新的市场，更容易地发现商业机会。通过在互联网上全天候地提供产品及服务的最新信息，商家可以与客户和消费者随时建立紧密联系来确保他们的竞争优势。互联网在网络世界为公司提供了大量的市场和产品推销机会，同时也使他们与顾客加强了联系。利用多媒体技术，可在互联网上非常容易地建立起法人形象、产品、企业商标名称。详细精确的销售数据有利于降低库存，节省运转费用。顾客的消费模式、个人爱好及购买能力等详细信息能帮助商家有效地调整营销策略。

在建立电子商务系统之前，必须确定以下问题：建立电子商务系统的目标是什么？公司在什么范围内采用电子商务？有多少投资？建立电子商务系统也必须考虑公司的硬件和软件，以及建立电子商务系统所必需的技术。

1．电子商务模型

电子商务区别于传统商务的特征是传送信息的手段和处理信息的方式。实现信息传送和处理的变革，无疑需要两种服务的支持：通信和数据管理。另外，为了使电子商务应用在一个可操作的设置上得到实际的应用，安全也是必需的。

电子商务模型必须包括通信、数据管理和安全。当然，应用程序也是模型必不可少的一部分。因此，一个电子商务模型主要由用户界面、应用程序、通信、数据管理和安全 5 部分构成。需注意的是，电子商务模型包括用户界面元件，表面上看，好像没有包括在上面的讨论中。这是因为用户界面和应用程序结合紧密，以至于经常被作为不可分割的整体。

因此要形成普遍的电子商务环境，就必须要求配置一个网络，这个网络有能力提供大量用户和服务提供商之间的连接。硬件、软件和安全问题都必须加以考虑。需要那些利用网络优势，提升计算机性能的新的应用。为了支持新的应用程序，需要那些能利用不断增长的带宽的网络软件。许多应用程序的发展将超乎想象，那些允许用户方便地利用这些新环境所提供服务的复杂用户接口将会出现。开发电子商务环境的关键组件之一将是那些能够支持本地应用程序又同时能和异类网络环境交互的软件，这要求使用那些构建在国内和国际标准之上的开放系统。

2．电子支付系统

一个电子支付系统（EPS）就是为了对购买由 Internet 提供的信息、货物和服务进行支付，如书、音乐光盘、软件、个人计算机、网站维护服务、研究报告等。它也是一种对在 Internet 以外所提供的商品和服务进行支付的便捷方式。由于支付有助于自动化销售活动，扩大潜在客户数量，因此电子支付提供了一种低花费、高效率的选择。

（1）智能卡

智能卡是如同信用卡大小的小型电子装置，其中包含电子存储器和一个嵌入式的集成电路（IC）。包含集成电路的智能卡有时被称作集成电路卡。使用智能卡，无论是从中提取信息还是向卡中加入数据，都需要一台智能卡读卡机——一台将智能卡插入其中的小设备。智能卡技术的优点之一是消费者可以将自动柜员机接到他的计算机上。现金可以很容易地通过计算机从银行账户转成智能卡中的数字代币。使用者可以使用这些数字代币向商家进行 Internet 在线支付。这种芯片到芯片的货币传输开辟了使用私有、公有密钥加密解密算法的安全机制的新纪元。

（2）信用卡

信用卡（借记卡共享其网络）是在线消费者支付的首选方案。信用卡成为一种支付方法而流行的原因有：易于使用，没有技术障碍，几乎每个人都有一张；消费者信赖发卡公司能够提供安全交易；大多数商人接受信用卡。在信用卡交易中的参与者有顾客、顾客的信用卡发行人（发行的银行）、商人、商人的银行、在 Internet 上处理信用卡事务的公司。

（3）电子现金

一个人有时需要购买像一张报纸这样的小额价值的商品。由于高额的交易费用，使用信用卡不适合在线购买这些商品。电子现金系统，如 E-cash，允许顾客将现金存放在一个银行账户中，则顾客将收到一个代表一个货币单位的编码数字串，该数字串被传送到用户的硬盘之中。接着，顾客可以将现金传递给在 Internet 上也支持该系统的商家。然后，商家通过将电子现金存入银行从而将这些数字现金兑换成实际的现金。

3．电子商务可能产生的影响

与传统商业相比，电子商务具有以下优势。

（1）覆盖范围广

一个结合互联网、内联网（企业内部的局域网）和外联网（企业外部网络）的网络系统使买方、卖方、制造商及其合作伙伴能够在全世界范围相互联系和方便地传输商业情报和文件。

（2）功能齐全

在电子商务中，不同类别和不同层次的用户能够实现不同的交易目标。例如，发布商业情报、在线谈判、电子付款、建立虚拟市场和在线银行等。

（3）使用方便灵活

基于互联网，电子商务不受专门数据交换协议的限制。可以使用任何类型的个人计算机，在世界上的任何地点，在计算机屏幕上方便地进行交易。

（4）成本低

利用电子商务，能够大大削减用于雇佣员工、维持仓库和店面的费用及国际旅行和邮寄的开支。使用互联网的费用非常低。

电子商务将对社会经济产生重大影响。

（1）电子商务将改变人们在商业活动中习惯采取的行为方式

通过网络，人们可以进入虚拟商店，到处浏览，挑选他们感兴趣的东西，以及享受各种在线服务。另一方面，商家可以通过网络与消费者联系，决定购买物品（种类和数量）并结账。政府机构能够通过网络进行电子招标和政府采购。

（2）电子商务的核心是人

它是一个社会系统。在线商店改变了人们的日常生活方式，充分体现消费者在交易中的自主权。

（3）电子商务改变企业生产产品的方式

通过网络，制造商直接了解市场需求，并且按照消费者的需要安排生产。

（4）电子商务极大地提高了贸易效率

电子商务能够消除中间环节，最大限度地降低销售成本。生产安排能够实现"小批量加品种多样"，以及"零库存"将成为现实。

（5）电子商务呼唤银行服务改革

消费者采购时再也不用实际现金了，像在线银行、在线现金卡和信用卡、在线结账、电子发票和电子"现金"将成为现实。

（6）电子商务将改变政府行为

电子商务也叫"在线政府"，它是一个在线行政管理机构。

阅读资料

商业网络技术

电子商务是一个系统，它不仅包括可以直接产生收益的那些商品交易，这些交易主要集中在货物买卖和服务上，而且也包括那些支持产生收益的交易，比如那些商品和服务所产生的需求，提供销售支持和客户服务，或提供商业伙伴之间的通信等。电子商务是在传统商业的优势和结构上，增加了计算机网络提供的种种灵活性而形成的。

电子商务产生了一些新的商业模式和一些新的运营方式。例如，亚马逊 Amazon.com 是一家在华盛顿州西雅图的图书经销商。这家公司没有实际的书店，而是通过互联网售书，与出版商合作直接发送图书。因为其所有产品（商业软件包）都是电子的，都能存储在同一台计算机中，这些计算机用于处理订单，并作为 Web 服务器使用，其库存则全部是数字化的。

信息技术使得新的企业运营方式成为可能，这就意味着会产生新的或改进了的产品和服务（如汽车和航空）、附加的销售渠道（例如互联网银行业务）、促成更有效的形式（如借助电子市场获得全球条件）、新的供应商和客户可以合作的方法（如合作计划）、新的服务（如虚拟社区）、高效管理（如对关键性能指标的自动测量）或新的信息服务（如产品目录）。

可以想象这样一个情景，每个雇员、每个客户、每个商业伙伴、每个设备和每台计算机都可以在任何时刻立即访问相互之间的数据。

网络使得通过协调可以从一个信息源向客户提供所有的产品和服务。不是客户，而是供应商才是处理汽车所有权方面的专家。汽车分销商的任务是帮助客户选择汽车、拿到检测报告、财务结算、驾车、转卖等。这同样适用于汽车商上游业务的几个阶段，例如，处理一辆汽车的日常业务，也可能包括在线查找早期的问题或访问信息提供者。信誉评价机构给出的不仅是客户的信誉度还提供延迟支付或贷款、担保等具体信息，轮胎分销商不仅提供轮胎，还提供CAD（计算机辅助设计）数据和测试报告以及批量质量数据等。

从企业观点看到的大量IT的开发，在企业变革中都负有以下7项使命。

- 企业资源计划（ERP），也就是企业的运作实施，总是悄悄地在后台运行。使管理上的综合应用、产品开发和技术更集中在商业上而不是在管理上成为可能。
- 知识管理为每项任务提供所需要的有关客户、竞争对手、产品以及上面的所有过程本身的知识。
- 智能设备把信息处理带到作业点：交通信息通过卫星导航系统（GPS）传给汽车驾驶人，生产厂家现金出纳机上的销售信息传递给产品生产厂家，而机器故障信息经由传感器传给服务工程师。
- 商业网络使两个公司之间的合作简单得能以一个企业的面貌出现。最终产品的销售信息立即可为供应链上所有公司使用。
- 一些公司在某时刻单独运作的很多子过程可从网上作为电子服务提供。例如顾客模型加工，除了供应商以外，一个第三方在线数据库提供者和顾客本人可以承担其模型加工责任并通过电子服务提供模型。
- 公司将不只简单地销售产品或服务，而是支持全客户流程。运输商务将使用物流处理，医生将支持整个治疗过程，保险公司将代替客户处理索赔流程。
- 合作管理将不再只关注财务结果，也关注产生这些结果的因素。财务管理将变为评价企业成功的关键性能指标的价值管理。

6.2　网站导航

课文

网站导航设计是关于链接的。它是在一个站点中决定页和内容重要性与相关性的重要因素。这需要决定在信息页间是否创建有意义的关联。同时，通过导航元素不仅能够找到人们正在找的信息，而且能够以不同的方式体验这些信息。

导航设计的共同目标是轻松地实现信息交互。导航对用户来说应该是"无形的"。然而测量它的效率则是有疑问的。当不注意它的时候，很难展现它的价值。同时，导航也是极度复杂的。要对成千上万的网页提供访问，在不同网页之间建立关联，以及在不计其数的流程图详细描绘的交互作用中都要使用链接。在任意一个页面的导航只是一个大型系统的一小部分，但有时候是很难掌握的。

当为你的网站设计导航的时候，你应该通知用户发生了什么，你所设计的导航系统为用户在整个网站漫游提供了导航的关键。文本以及标签是人们知道当前页是什么或者哪个标题的主要导航方式。但是，除此之外，导航反馈的方式有两种：在选择导航之前滚动；在过渡到新页之后显示位置。

在导航中，通过高亮显示目录来显示页在网站中的位置能够帮助访问者进行定位。在信息资源丰富的站点，定位是有价值的。如果某人被打断了进程需要恢复到刚才的那个页面，显示定位也是有帮助的。

在设计网页导航之前，需要花时间调查人们是怎样查找信息的。你考虑得越多，在解决设计问题时，就越容易构建你的想法。网页导航可以从以下三个方面定义。

（1）人们如何从网站的一个页转到另一个页的理论与实践。

（2）直接寻找目标和定位超级链接信息的过程：浏览网页。

（3）所有的链接、标签和其他元素当在特定网站互动的时候都提供存取页，并且帮助人们在网页上定位。

链接是连接到另一个页面或者本页内不同地点的文本或者图形。它们允许从一个想法到另一个想法的可能的飞跃。例如你正在读关于中国的外交政策的信息，你能跳转到关于国家人口统计详细信息的页面，这就是链接的功劳。但是，链接能做的工作不仅仅是从一个页面跳转到另一个页面，它们也能显示重要性，展示相关性。

在我们设计网页的经历中，我们都知道导航在其中占有重要的地位。它在一定程度上以增强理解、折射品牌和提高整个网站的可信度的方式提供访问数据的方法。因而，网页的导航能力和查找信息的能力对使用者来说有很大的经济影响力。

网页导航设计不仅仅限定在选择一排按钮的任务上。它的范围更宽广，同时，也更精巧。网页设计者要协调用户和商家的目标。这需要对各种信息的理解，如应该具有信息组织、页面布局以及设计表达方面的广泛的知识。

总之，要将各种各样的信息汇集起来创建导航系统。尽管访问者会感觉这个系统如同一个整体，但是我们可以剖析它们的各个组成部分。例如，如果标签在页的顶部中心（以首页为开始），那么它就是主要的导航。

这是一个好的导航系统吗？答案也许是相对的。你务必考虑多种因素，从企业目标到用户目标。但是，我们有评价导航系统好或者坏的共同原则。例如，在 BBC 页面上，导航是平衡的和一致的，并且给你一个清晰的指示，你在哪里。总之，为此类站点设计的导航是恰当的，可能"让人们知道想去哪里"是评价导航设计是否成功的最重要因素。

以下列出了一些成功的导航系统的重要性质，它们不是规则，不规定怎样设计导航系统，但是了解它们可以让你在设计时有指导的方法。总之，这些方面可以预测导航系统的效率。

- 平衡性。
- 易学性。
- 一致性（不一致性）。
- 反馈性。
- 有效性。
- 清楚标签。
- 视觉清晰性。
- 站点类型的适应性。
- 用户目标的排列。

在网站上，广度和深度是关于信息结构和导航系统的函数。例如，在一个网页上经常显示两个层次的目录，因此会减少二级页的点击数。动态菜单有一个相似的作用，使用它们可以直接在网站上从顶层页面访问更深层次的页面。

通常，更宽的结构比更深的层次要好。在深层次和多层次上进行的选择所付出的努力超过在更广领域上进行的选择。眼神比单击鼠标（装载页）快捷。尽管用户更倾向于深度结构较快些，不倾向于广度。一直显示所有链接将使选择变得更难。用户也许就选择使用他们需要的第一个选项，或者简单地放弃。

通往信息的路径应该是高效率的。应该努力创建易发现易单击的导航链接、表和图标。例如，需要手和眼协调一致的动态菜单作为动态导航会降低用户的使用速度，而与你创建的链接、按钮、表和菜单之间的交互却是需要付出小小的努力就能完成的。

导航设计的共同的战术就是成群地选择，相似的一起编组，提供焦点层。用户不必扫描网页上面的每个链接，他们可以看部分标题，在那里他们可以关注于该区域的链接。这是一个两阶段扫描过程：首先找到正确的组，然后放大单个的链接。

人们不期待需要学会如何使用网站。也没有训练期或者必要的学习指南或者一组指令集。在开放式网页上，在单页上停留的时间可以典型地以秒来计算。导航的意向和功能一定是迅捷且清晰的，特别是那些有企业目标的资源丰富的站点，这也对任何类型的网页导航有用。

"一致"是交互设计的基本原则。至于导航，通常是指在页上稳定的位置中的出现机制或者链接，具有可预见性的表现形式，有标准的标签，且整个网站看起来都相似。通常，这是你应该力争使用的好办法。但是要记住的是一致不等于是均一，不一致恰恰也是导航设计中很重要的。以不同方式运作的事情也应该以不同的方式表现。当人们说"要一致"的时候，真正的规则是：在一致与不一致间保持平衡。某些不一致对导航来说是至关重要的。通过在网站中改变位置、颜色、标签或者通用布局能创造一种进展感觉。

阅读资料

社交网络

社交网络是指将一群个体组成特定的团体，就像是小的农村社区或者社区分部那样。尽管社交网络是可以亲身参与的，特别是在工作单位、大学或者高中，但它在线上更加常见。

这是因为，与大部分高中、大学或者工作单位不同，互联网上有着数以百万计的人。他们正在寻求与他人相识，以获取或分享关于烹饪、打高尔夫、园艺、发展互助职业联盟、求职、企业对企业营销，甚至有关针对"发展运动"烘制饼干的第一手信息和经验。这些话题和兴趣点就像世界上的趣闻一样丰富而多彩。

提到在线社交网络，常用的是网站，这些网站就称为社交网站。社交网站的功能就像是一个互联网用户的在线社区。凭借被谈论着的网站，这些在线社区的许多会员可以分享在爱好、宗教、政治以及另类生活方式方面的共同兴趣。一旦你被准许访问社交网站，你就可以进行社交活动了。这种社交活动也许包括阅读其他会员的人物简介页面，甚至有可能联系他们。

交友只是在线社交网络的诸多益处之一。社交网络的另一个好处就是多样性，因为互联网允许世界各地的人们都可以访问社交网络。这就意味着尽管你在美国，你同样可以与在丹麦或印度的某个人发展在线友谊。你不但能够交到新的朋友，而且或许可以学到有关新的文化或新的语言的不少东西。学习总是一件好事。

正如所提到的，社交网络通常专注于将特定的个人或机构一起组成团体。尽管有一些社交网站有着特定的兴趣方向，但其他的则不是这样。那些没有特定兴趣方向的网站，通常被称为"传统的"社交网站，并且一般都开放了会员制度。这就意味着任何人都可以成为会员，无论他们的爱好、信仰或观点是什么。然而，一旦你处于这个在线社区中，你就可以开始经营自己的朋友圈，并且清除那些与你兴趣爱好、目标不同的会员。

众所周知，存在一些与社交网络相关的危险，包括数据盗窃和病毒感染，这些危险呈上升趋势。不过，最为普遍的危险往往包括在线"掠夺者"或者虚假身份。虽然这些危险确实存在于在线网络上，但它们也同时存在于真实世界。正像你在俱乐部和酒吧、学校，或者工作时遇到陌生人时被劝告一样，你也同样被劝告在线上要谨慎行事。

通过了解自己的网络环境和交谈对象，你应该能够安全地享受在线社交网络。可能需要进行许多的电话交谈才能了解一个人，但是直到能够亲自见面，你确实无法给对方一个明确的评价。

如果你消息灵通，并且满足于自己的发现，你就可以开始从数百个网络社区当中搜索以便加入。通过进行标准的互联网搜索也很容易做到这一点。你的搜索有可能会返回多个结果，包括 MySpace、FriendWise、FriendFinder、Yahoo! 360、Facebook、Orkut 以及 Classmates。

6.3 移动商务

课文

随着万维网的引入，电子商务彻底改变了传统的商务，大大地推动了销售额以及商品和信息的交换。最近，无线网络和移动网络的出现使得电子商务进入了新的应用及研究主题：移动商务。移动商务可以定义为：通过移动的无线电通信网络进行的涉及货币价值的任何交易。一个不太精确的说法是把移动商务描述为人们通过连接网络的移动设备所获取一系列应用及服务的兴起。移动商务可以有效、方便、随时随地给客户传递电子商务。很多大公司已经意识到移动商务带来的好处，开始为客户提供移动商务选择。

从拥有智能手机或平板电脑到搜索产品或相关服务，浏览这些信息，最后下单购买，这只是几个简单的步骤。结果就是移动商务的业务一年的增长超过了 50%，远高于桌面电子商务的业务一年 12% 的增长率。移动商务的高增长率当然不会一直持续下去，但分析人士估计到 2017 年，移动商务将会占据整个电子商务的 18%。

一份对排名前 400 的移动公司销售的研究表明，在移动商务中，零售物品占了 73%，旅游业务占了 25%，同时还有 2% 是票务。越来越多的顾客正在利用其移动设备搜索人物、地点以及餐馆，或他们在零售店看过的商品的交易。顾客从桌面平台到移动设备的快速转换驱动着移动营销花费的上涨。

与电子商务系统相比，移动商务系统要复杂得多，因为它必须包含与移动计算相关的成分。下面简略地描述了一个典型的由移动用户提交请求后所引起的处理过程。

- 移动商务应用：内容提供商通过提供两套程序实施应用。客户端程序，如微浏览器上的用户界面；服务器端程序，如数据库访问及更新。
- 移动站：移动站向终端用户显示用户界面，这些用户在界面上细化他们的要求。然后，移动站把用户的要求传递到其他组件并在用户界面显示处理的结果。
- 移动中间件：移动中间件的主要目的是不留痕迹地、清晰地把互联网的内容绘制到移动站上，该站支持各种操作系统、标记语言、微浏览器及协议。大多数移动中间件也通过加密信息在一定程度上保证交易的安全性。
- 无线网络：移动商务之所以可行的很大原因是无线网络的应用。用户的要求被传递到最近的无线接入点（在无线局域网环境里）或者一个基站（在蜂窝网络环境里）。
- 有线网络：对于移动商务系统来说，该部分是可选的。

- 主机：该部分同电子商务所使用的这部分一样。用户的要求通常在这里得到执行。

没有信任与安全就不会有移动商务时代。在客户连接支付业务或用其移动设备进行购物时，如果供货商或支付提供者不能给客户安全感，那怎么能指望吸引客户呢？对消费者来说他们需要对以下事情感觉舒心：他们不会被要求为他们没有使用的服务付费；他们的支付细节不会落在别人手里；有现成的适当的机制来帮助解决纠纷。

当我们在评论移动安全性的各个方面时，应该牢记的是安全性总是需要全面考虑的。一个系统的安全性只相对于它最弱的部分，而保证网络传输只是安全的一部分。从技术方面来说，网络安全涉及到很多不同的方面，每个方面都对应着受到的威胁或薄弱环节的一个层面。防御其中一个层面不能保证你不受其他层面的侵扰。

移动商务应用广泛。下面列出了主要的移动商务应用以及每一类的详细情况。

- 商务

商务是指大批货物的买卖或者交换，涉及从一处到另一处的运输。移动商务技术的便利与无处不在推动了商务的发展。有很多例子表明移动商务是如何促进商务的。例如，消费者现在可以用他们的便携电话为自动售货机里的商品付款或者支付停车费；移动用户不必去银行或自动取款机就可以查看他们的银行账户并进行余额转账等。

- 教育

由于缺乏计算机实验室空间、教室和实验室分离、为了装有线网络对旧教室进行改造等问题，很多院校都面临着困境。为了解决这些问题，无线局域网通常用来将计算机或移动手持设备连接到互联网或其他系统。因此，学生不必去实验室就能获取很多必要的资源。

- 娱乐

娱乐一直在互联网应用上起着十分重要的作用，对于更年轻的一代来说也许是最广泛的应用。移动商务使得随时随地下载游戏、图像、音乐、视频文件成为可能，也使在线游戏和赌博更容易进行。

- 卫生保健

卫生保健费用昂贵，而移动商务可以帮助减少费用。通过移动商务技术，医生和护士可以立刻远程获取病人记录并将其更新，而在过去，很多事情都是因为这项工作不能及时进行而耽搁。这改进了效率和生产率，减少了行政开支并提高了总的服务质量。

- 库存跟踪及发货

及时的投递对今天的企业取得成功至关重要。移动商务让企业可以跟踪其移动库存并适时地投递，这样可以改善客户服务、减少库存并增强公司的竞争力。大多数投递服务公司如"联合包裹服务公司"和"联邦快递"都已经把这些技术应用到其全球范围的企业运作中。

- 交通

交通管制对于很多大城市来说是一件头疼的事。用移动商务技术可以在很多方面轻易改善交通。例如，移动手持设备预计会拥有全球定位功能，如确定司机的位置、指示方向、对该区域的交通现状进行咨询；交通控制中心会根据车里移动设备发出的信号监控并管制交通。

- 旅行与票务

旅行费用对企业来说可能是非常高的。移动商务通过向商务旅行者提供移动旅行管理服务可以帮助减少经营成本。

阅读资料

物联网

物联网建立在现有技术的基础上，例如，射频识别技术（RFID），并且随着低成本的传感器的可用性，数据存储的费用的降低，可以处理万亿数据的"大数据"分析软件的发展，以及允许为所有这些新设备分配互联网地址的 IPv6 技术的实现，使物联网正在成为可能。

物联网有时也被称为物体的互联网，将会改变包括我们自己在内的一切。这虽然好像是一个大胆的声明，然而考虑到互联网已经对教育、通信、商业、科学、政府和人类发生的影响，显然互联网是人类历史上最重要和强大的创造之一。

从技术角度来看，物联网不是单个新技术的结果，相反，它是几个互补技术的发展共同提供的功能，这些功能组合在一起有助于弥合虚拟世界和物理世界之间的差距。这些功能包括：

- 沟通与合作：物体拥有将互联网资源连成网络，甚至实现物体间相互连接，以及使用数据和服务，并且更新物体状态的能力。在此，无线技术，如 GSM、UMTS、无线网络、蓝牙、ZigBee 以及其他各种无线网络标准都与此紧密关联。
- 寻址能力：在物联网内，物体可以被定位，并通过发现、查找或者命名服务进行寻址，因此可以远程查询或配置。
- 识别：物体是唯一可识别的。RFID、NFC（近场通信）和光可读条形码都是这类技术的应用实例。采用这些技术后即使没有内置能源的被动物体都可以被识别（借助于如 RFID 阅读器或移动手机这样的"中介"）。如果"中介"被接入网络，则识别功能就可使物体被链接到与该特定物体相关联的信息，该信息还可以由服务器检索到。
- 感知：物体利用传感器收集其周围的环境信息，记录、转发该信息，或者直接对该信息做出反应。
- 执行：物体包含执行机构（如将电信号转换成机械运动）来操纵其周围环境。这种执行机构可以通过互联网远程控制真实的过程。
- 嵌入式信息处理：智能物体带有处理器或微控制器，以及存储空间。例如，这些资源可被用于处理和解释传感器信息，或者向物体中存入有关物体是如何被使用的信息。
- 本地化：智能物体能知道其物理位置，或者被定位。全球定位系统或移动电话网络是适于实现该功能的技术，另外还有超声时间测量，UWB（超宽带），无线电信标（例如，相邻的 WLAN 基站或具有已知坐标的 RFID 阅读器）和光学技术。
- 用户界面：智能物体能以适当方式与人通信（不论直接或间接，如通过智能手机）。与此有关的创新互动方式有诸如可触摸的用户界面，基于聚合特性的柔性显示，以及语音、图像或手势的识别方法。

最为特定的应用程序只需这些功能的一个子集，特别是由于实现所有上述功能通常很昂贵，并且需要复杂的技术。例如在后勤应用中，目前集中在近似定位功能（即上一次读取点的位置）并对使用 RFID 或条形码的物体进行相对低成本的识别，传感器数据（如用于监视低温运输）或嵌入式处理器只被用于不可或缺这些信息的物流应用中，例如有温度控制的疫苗运输。

就像物联网听起来的那样酷，它也存在缺点。隐私性、可靠性以及数据控制方面的问题仍然亟待解决。

6.4 电子支付与物流

课文

电子支付

在电子商务过程中，支付是非常重要的一个环节。网上支付意味着以电子方式处理资金转移、收入或付出。网上支付，有时称为电子支付，是应电子商务应用的要求而发展起来的。有许多种数字资金或电子货币，它们主要可以分两类：一类是实时付款式的，例如电子现金、电子钱包和智能卡等；另一个是事后付款机制，例如，电子支票（网上支票）或电子信用卡（万维网信用卡）。

先进的加密与确认系统能使数字现金使用起来比真纸币更安全和隐私，与此同时，还保留了纸币的"匿名支付"的特点。数字现金系统是建立在数字签名和加密技术之上的，这个系统常利用公钥/私钥密钥对方法。

首先，用户存钱到银行，开个账户。银行将给客户可用于客户端的数字现金软件和一个银行的公共密钥，银行用服务器端数字钱币软件和私人密钥对信息加密，用户可用客户端数字钱币软件和银行的公共密钥将信息解密。当客户需要一些数字现金支付某人时，就用数字现金软件生成他所需要的数字现金，给它加密并传给某人。某人将这个信息再加上他自己的账户信息和存款指令一起加密并且发送给银行。银行用私人密钥解密信息并证明客户与某人的身份，钱从客户账户上转至某人账户。

由于数字现金只是一组数字，所以需要确认其真实性。银行必须具备能追踪电子现金是否被复制或重复使用的能力，而又不能将个人购买行为与从银行购现金的人联系起来，以保持真钱币的"匿名"特色。在整个过程中，商家也不必知道任何客户的个人资料，它可用银行的公共密钥检验客户送来的数字现金。

另一种网上支付方法是电子支票，它的前身是用在增值网中的财务 EDI 系统。有些电子支票软件可装置到智能卡上，这样，客户能非常方便地携带它而且可用在任何计算机上写电子支票。一些电子支票软件可为客户提供电子钱包，使他们在没有银行账户的情况下也可以写支票。

电子银行不是真正的银行，它是一种虚拟银行，在这个虚拟银行中，银行家、软件设计者、硬件生产者和 ISP 们共同合作，完成各自领域的工作并很好地与各方协调。所有服务都在银行服务器上被处理，银行通过网站与客户互动。客户也可以用便携式智能卡在任意一台联网计算机上来进行银行支付，例如，付账、购买电子货币及注册网上银行等。

这些新型银行服务软件采用了现代信息技术，能提供方便、实时、安全和可靠的银行服务。此外，它们还可以提供其他相关服务，例如股票行情、税收方法及政策咨询、外汇汇率和利息率信息。软件越方便，则顾客越多，利润也可更多，服务则可更好。这样，银行和客户为双赢关系，客户获得了便宜而方便的服务。银行极大地降低了运作成本。因为在网上，经营一个网站的开支远远比在地上建一个银行分支便宜。而且，在网上，服务一个人的成本与服务一万个人的成本没有什么不同。随着网上银行的飞速发展，软件公司和网络提供商得到更多业务，它们都与银行处于双赢关系。

物流

正如商品的流动，信息也必须是流动的。在合适的时间，合适的条件下，携正确的文件把商品送到正确的地点，这些有关"正确"的问题必须要弄清楚。物流中信息系统的本质就是把准确的数据转换成有用的信息。错误的数据和匮乏的信息就会扰乱物流管理活动。当然，即使是有了精确的数据和丰富的信息，还必须有人付诸实际行动。

顶级的物流效率和效果要求有一个很好的集成物流信息系统（ILIS）。没有随时能够访问的准确信息，集成的物流管理操作就会失去效率和效果。集成的物流便不能维持战略竞争力。ILIS 的优先应用领域包括管理库存状态，追踪货品和发货，取货和运输，订货便利化，订货准确化，内部物流和外部物流的协调，以及订单处理。通过 ILIS 信息流的质量是至关重要的。在信息的质量上，有三点值得我们关注：（1）获得正确的信息；（2）保持信息的准确性；（3）有效地沟通信息。

一个集成的物流信息系统可以被定义为：通过人员、设备和一定流程，将所需信息加以收集、整理、分析、评估，然后将它们以及时、准确的方式发给恰当的决策人以帮助他们做出高质量的物流决策。

ILIS 收集来自所有可能渠道的信息，以协助集成物流经理做出决策。它接触到市场、金融和制造业信息系统。所有这些信息都将被高层管理者用来制定战略决策。

ILIS 有四个主要的组成部分：订单处理系统，研究和智能系统，决策支持系统及报告和输出系统。上述四个子系统应该共同提供给集成物流经理作为及时准确决策参考的依据。这些子系统与集成物流管理功能和集成物流管理环境相连接。在开发信息之前，信息需求就必须确定下来。同样，一旦产生基于需求评估的信息，它将被送到集成物流经理那里。

订单处理系统无疑是最重要的子系统。订单处理是一系列使正确的货物得以准备好并运送到客户（直到库房接货为止）的活动。订单处理包括检查客户信用，抵补销售代表的账户，确保产品的供应，并准备必要的货运文件。卖方应能控制订单周期活动。通过计算机的应用，订单处理的时间已经大大地缩短了。

研究和情报系统（RIS）不断地监控环境，观察并总结影响整合物流操作的事件。RIS 监控公司内部环境、外部环境和公司之间的环境。外部环境包括在公司之外发生的、通常不在公司控制下的事件。公司间环境包括一些直接影响公司，且公司有一定控制的外部环境要素，如分销渠道。公司内部环境包括公司的内部工作和被公司掌控的要素。

决策支持系统（DSS）以计算机为基础，运用分析建模来解决复杂的集成物流问题。所有 DSS 的核心是一个包罗万象的数据库，包含能使集成物流经理用来做决策的信息。

ILIS 最后的子系统是报告和输出系统。常规的报告用来制定计划、操作和控制整合的物流。计划输出包括销售趋势、经济预测和其他的市场信息。营运报告用于库存控制、运输调度、发送、购买和生产计划安排。控制报告用来分析费用，预算和业绩。

阅读资料

电子营销

通过互联网，厂商可以不利用中介，直接接触客户。只要制造商销售的是已经确立的品牌而且他们的主页已经有了很好的知名度，制造商的直销就可以实现。如果一个制造商的网站知名度不高，仅仅开办一个主页，被动地等待顾客点击，这个也许对销售贡献不大。因此，公司有必要大张旗鼓地宣传他们的网站地址。任何经济上划算的广告都可以被用来实现这一目的。其中一个例子是，把自己的网站链接到知名的电子目录网站，而且大多数厂商都使用中介机构的目录服务。这些中介网站被称为电子购物中心。

起初，电子营销主要关心的是安全技术，这些技术对于基于互联网的营销是必要的，如强大的搜索能力和安全电子支付。然而，在今天，管理的重点正转移到如何利用基于互联网的营销机会与现有销售渠道一起来提升竞争力。

越来越多的公司将其客户分成不同的客户群，并为每个客户群提供不同的针对性的信息。在有些情况下，当公司采用网络时，这些目标群体的规模可能更小，有时候甚至可能一个客户就是一个目标群。网站访客行为的最新研究甚至标明，网站可以采取一些方法来响应不同的时间、有不同需求的访问者。

绝大多数的公司使用术语"营销组合"来描述他们用以实现其销售、促进其产品和服务目标的元素的组合。当一家公司决定使用某些元素的时候，它便把那个特定的营销组合称作它的营销战略。即使那些在同一行业的公司，他们都努力地在市场中表现出与众不同。公司的营销策略是一个与现有客户和潜在客户能获得公司信息的网上展示相结合的重要工具。

大多数的营销课程把营销的基本问题归纳成 4P，即产品、价格、促销和场所。产品是一家公司正在出售的物理实体或服务。产品的本质特性很重要，但是客户对产品的感知和产品的实际特性一样重要，客户对产品的认知度被称为"品牌"。

"营销组合"的价格元素是客户为获得产品所付出的资金数量。近几年来，不少营销专家认为公司应该在一个较宽泛的意义中来考虑价格因素；也就是说，价格应该是客户为获得产品所付出所有费用的总和。从该总成本中减去客户从所获得的产品中得到的收益就能产生客户从交易中获得的客户价值的估计值。Web 为创造性定价提供了新机遇，也为在线拍卖、反向拍卖及群体购买战略的价格协商创造了新的机遇。这些基于网络的机会正在帮助公司发现新的增加客户价值的方法。

促销包括发布有关产品消息的任何方法。在互联网上，极大地推广可能性存在于与现有客户和潜在客户的交流之中。公司已经使用互联网，通过电子邮件及其他的方法与他们的客户进行有意义的交流。

当公司通过网络开展其商业活动的时候，应该能够创造一个有足够柔性的网站来应对不同访问者的需求。公司应该建立能够满足不同类型客户特定需求的站点，而不是仅仅建一个产品展示的站点。制定一个基于客户的良好的营销策略的第一步就是要甄别拥有共同特性的客户群体。在 B2B 网站上，基于客户的营销方法被大力倡导。相对于 B2C 网站的运营者，B2B 卖家更清楚知道，要满足他们客户的需求，就必须客户化他们的产品和服务。近几年来，B2C 站点逐渐地将以客户为基础的营销元素加入了他们的网站之中。

做广告就是为了影响买卖双方的交易而进行的信息传播的努力。互联网重新定义了广告的含义。互联网使得消费者能直接与广告商及广告互动。互联网为广告商提供了双向通信和电子邮件的能力，同时使得广告商把广告费花到他们想针对的特定群体身上，这比传统电话营销更准确。最终互联网实现了真正的一对一广告。

第 7 章 计算机应用

7.1 办公自动化

课文

办公自动化（OA）是应用计算机和通信技术去改善办公人员和管理人员工作效率的技术。在 20 世纪 50 年代中期，这一术语就被用作各种数据处理，主要是文书工作自动化的同义词。经若干年搁置后，20 世纪 70 年代中期该词再次被用来描述字和文本处理系统的交互使用，这种系统后来又与强有力的计算机结合导致所谓的"未来综合电子办公室"的出现。

基于个人计算机的办公自动化软件在许多国家已经成为电子管理不可缺少的组成部分。文字处理程序取代了打字机；电子表格取代了账簿；数据库程序取代了传统的纸选票，库存品和职员列表；个人管理程序取代了纸日记簿等。

办公自动化包括六种主要技术：

- 数据处理——通常指由计算机计算的数值型信息。
- 字处理——文本型信息，即文字与数字。
- 图形——可以是数字和文字形式信息，将其从键盘输入到计算机，再以图形、图表、表格或者更易于理解的视觉直观形式显示在屏幕上。
- 图像——画面形式的信息。取实际画面或照片输入到计算机并在屏幕上显示。
- 声音——语音形式的信息处理。
- 网络——用电子装置把计算机和其他用于处理数据、文字、图形、图像和声音的办公室设备连接在一起。

主要生产厂家最初出售的系统是针对办公人员和秘书的，这些设备主要做字处理和记录处理用（保存像名字和地址那样的小型顺序文件，最后经分类和合并形成信函）。现在，注意力已集中在直接为经理和业务人员提供办公自动化系统上了（见图 7-1），这些系统强调管理通信功能。

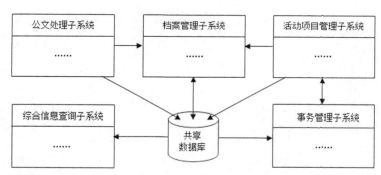

图 7-1　办公自动化系统

当今的机构配置了各种各样的办公自动化硬件和软件，包括电话及计算机系统、电子邮件、字处理器、桌面印刷系统、数据库管理系统、双向电缆电视、办公室间的卫星广播、在线数据库服务、声音识别及合成系统。这些设备都力图使目前手工完成的任务和功能自动化。专家认为达到办公自动化的关键在于综合性——将各硬软件紧密结合成一个完整的系统，使得信息处理和通信技术应用最充分，人的干预最少。

现代商业活动的节奏急切地需求关键性的信息。与此同时，官方和商家也需要大量的文档。因此，现代商业办公室都在对传统的办公方法进行研究与调整，以找到可以随时随地获取信息和传递信息的较好的途径。这些部门正在寻求用来产生、记录、处理、归档、交流或发布信息的最有效方法。现代技术把办公自动化作为一个经济解决方案的基础。现在和将来的办公室系统都将致力于开发集成信息处理网络，这种网络能把一个公司日常处理的业务有效地汇集在一起。

（1）文字处理

文字处理是指使用计算机创建、编辑、打印文档的方法和程序。大多数标准的字处理程序具有以下特点：支持脚注和邮件合并功能，但不支持表格或列。界面采用可定制的工具条，并且编辑屏采用可缩放的草稿模式，可选择性地显示标题、脚注和页脚。不可编辑的打印预览方式显示整个页面或页的外观。当你选择一种新语言时，字体、键盘布局和输入方向均可变化，并且键盘布局可以定制。

文字处理软件各不相同，不过，所有的文字处理软件都支持一些基本功能。只支持这些基本功能（可能还有其他一些功能）的文字处理软件称作文本编辑程序。不过，大多数文字处理软件还有另外一些功能，以便于用户对文档进行更复杂的操作及格式化。这些高档的文字处理软件有时也称作全功能文字处理器，它们通常具有以下功能：

- 文本插入与删除
- 剪切、粘贴和复制
- 页面尺寸和边距调整
- 查找和替换
- 自动换行
- 打印
- 文件管理
- 字体说明
- 脚注和交叉引用
- 图形
- 页眉、页脚及页码
- 版式
- 宏
- 合并
- 拼写检查
- 内容及索引表
- 词库
- 多窗口
- WYSIWYG（所见即所得）

（2）电子表格

Excel 是一个电子表格程序，用于组织数据、完成计算、做出决策、将数据图表化、生成专业水准的报告、在 Web 上发布组织好的数据以及在 Web 站点上存取实时数据。使用 Excel，你可以在工作表上创建一个超链接，用来在本地网络上，即本单位的企业网上，或互联网上访问其他的 Office 文档。

电子表格可以用于各行各业，例如，会计可以用电子表格核对财务报表和编制工资表；在商业上，电子表格可用来编制预算和对报价进行比较；教师们用电子表格记录学生的考试成绩；科学工作者利用电子表格分析试验数据。家庭主妇利用电子表格记录家庭开支。

在 Excel 中，公式和函数用来对电子表格中的数据进行统计和计算，当数据改变时，计算的结果会自动地更新。所有的公式都以=号开始，它包含算术操作符、文字操作符、比较操作符和引用操作符——总共四种操作符。函数是在 Excel 中预定义的内置功能，包括 SUM、AVERAGE、COUNT 和 MAX 等。

电子表格中的各种统计图表示单元块中的数据，使这些数据简单且直观、易于理解；同时，当数据改变时，这些图自动地跟着变。

在 Excel 中，图包括"嵌入图"和"独立图"。前者与工作表一起放置、显示和打印，后者是独立产生的工作表，它可以独立于原来的数据表，分开打印。在 Excel 中有二维图和三维图。利用工具栏中的"图向导"或"插入"菜单中的"图画"命令按钮，它们可以逐渐地产生。用户可以编辑这些图，有折线图、条形图、饼图和其他的图供选用。

Excel 不仅能做简单的数据管理和比较，还可以建立数据库。利用数据表，可以增加或删除记录，并做排序、筛选、分类和汇总数据的工作。

（3）视频会议

视频会议是将远程地点通过单向或双向电视相连接。如果会议室内能安装所需的声像设备，通过举行远距离会议可代替面对面的会议，可节约旅行所花费的时间和经费。许多商业公司正试验通过视频会议召开销售和部门会议。这种花费仍然很高，尤其是当视频会议使用直接双向卫星信道连接时。

阅读资料

三维打印

你可能从新闻播报员和记者那里听说过三维（3D）打印并惊讶于他们的所见所闻。这种机器让人联想到星际迷航中可以凭空创造物体的复制机。它可以在玻璃、金属、尼龙以及超过百种的其他材料上进行"打印"。它可以用于制作无意义的小模型，例如被套印的尤达，同时它也可以打印制造的原型、终端用户产品、准合法的枪支、飞机发动机的零件，甚至使用人们自己细胞的人类器官。

我们生活在一个可以目睹许多人称为"第三次工业革命"的时代。3D 打印，或者被更专业地称为附加制造，让我们摆脱了亨利·福特时期的大规模生产线，并进入到新的可定制的、一次性生产的现实。

当用户需要洗衣机的某个零件时，在这个时代，他所订购的零件或许来自于一个用非常昂贵的模具注塑而成的一次性批量生产的零件，从中国运至分销商，分销商再交给修理工，最后到达他的手中。在未来，基于现已存在的技术，用户可以就在自己的家中仅仅用自己下载的 CAD 文件 3D 打印这个零件。如果用户没有合适的打印机，他也可以去当地的代工厂来打印。

3D 打印机使用了各种各样不同类型的附加制造技术，但是它们都有一个核心的共同点：它们会通过逐层地建立物体来创建三维物体，直至整个物体完成。这就特别像是在一张纸上进行二维打印，只不过是增加了第三个维度：高度，即 Z 轴。

这些打印出的每一层都是实际物体切成薄片的水平截面。试想一个多层的蛋糕，面包师每次放置一层直到整个蛋糕制作完成。3D 打印也有点类似，只不过它比制作蛋糕的过程更加精细一些。

尽管大多数人刚刚听说 3D 打印这个术语，实际上这项工艺流程已经被投入使用长达数十年。制造商在设计过程中，已经长期使用这种打印机来创建传统制造业的原型。但是直到最近的几年，设备价格一直昂贵并且速度不快。

现在，快速的 3D 打印机能够以数万美元的价格获得，并最终在制作原型的过程中，能够为公司节省数倍的花费。举例而言，耐克公司使用了 3D 打印机来创建多种颜色的鞋的原型。他们曾经在某个原型上花费数千美元以及数周的时间。现在，他们只需要花费数百美元，并且可以通过电脑在同一天的时间内实现对原型的即刻更改和重新打印。

一些公司正在使用 3D 打印机来满足短期制造或定制的制造，在这些情况下打印出的物体不是原型而是真正的最终用户产品。由于 3D 打印速度的提升以及成本的下降,人们希望得到越来越多的 3D 打印,并且能够期望更多的个性化的客户定制产品可用。

即使用户不能设计自己的 3D 模型,他仍然可以打印非常有意思的物件。用户可以从诸如 Thingiverse、3D 部件数据库,以及 3D 仓库等模型仓库中免费下载它们的模型文件。

人们都可以打印什么东西?这是没有限制的。一些人打印珠宝首饰之类的东西,一些人打印例如洗碗机等设备的备用件,一些人发明各种原创的东西,一些人创造艺术品,还有些人为他们的孩子制作玩具。当类如金属、塑料、玻璃,以及其他包括金银在内的材料可用时,几乎任何的东西都可以打印。

这是一个能够影响能源利用、消耗、定制、产品可用性、艺术、医药、建筑、科学,当然还有制造等方面的巨大的颠覆性技术。正如我们所知,它将会改变这个世界。

7.2 远程教育

课文

将计算机技术用于教育改善了几乎所有其他远程教育的方式。计算机技术作为教育环境下最大改变的发起者,联系和控制其他技术的能力使它成为所有技术的最前沿。计算机技术快速发展促使旧的设备更快地淘汰,经济现状是随着技术的发展价格不断降低。在教育领域有一点必须指出计算机经常被称赞为具有创新性和生产性的工具的原因,应归功于那些不断在编程中用大量时间同专家们合作的教育策划者。同其他媒体一样,计算机也存在弊端,但它们提供了缓解这些内在弊端的机会。计算机只是一种工具,也许并不是非常强大有利,但教育者可以用来创建使学习可以进行的教育环境。

计算机媒体通信的定义似乎与远程教育的定义类似,在远程教育中教师和学生在时间与方位上往往是分离的。计算机技术伴随着电子通信技术在信息存储和检索科学取得了巨大的进展,从而产生了重要的教育工具:互联网、远程通信、电子公告板、电子邮件、视频会议等。

视频光盘和激光技术的发展提供了一些独一无二的特点:存储大量数据的能力,无需加装光盘而显示静止图像的能力,在毫秒级内存取帧的能力。几乎瞬间的帧存取和巨大的静态存储能力使视频光盘成为独特的理想的视觉存储模式,它可以很方便地由教育者使用并由计算机控制,提供了一种可以用于所有教育形式中的交互媒体。

1. 互联网与远程教育

最新的发展和不断降低的费用提供了通过互联网从全球获得大量信息的途径。在世界任何地方的学生要获得信息所需要的只是一台计算机、一个调制解调器、电话和访问端口(商业的或教育的)。学生在家里就可以进行研究的功能正改变着教育机构研究的方式。电子出版领域是发展最快的领域之一,但是在它像其他印刷物一样被广泛地接受之前需要制定一些标准。

远程通信是通过提供传递系统能发送教育节目并使参与者之间互相沟通支持远程教育的。卫星和光缆用来发送教育节目为全国各地的学校提供了经济便捷的途径。像杰森项目和太空探索者一类的节目就可以使任何学校跟踪访问远程学习项目,世界范围内都能看到,并且与远程站点交流。光学技术的快速发展为传输日益增加的大量信息提供了方便,并且使相隔很远的学生通过电缆或电话连接来使用可以互相沟通的多媒体程序。

电子公告板及新闻服务为在世界范围内讨论各种议题提供了途径。这些"一对多"的通信平台使远程教育能够发送作业和课程信息。成组讨论使每个人能够就任何议题分析他们同学以及不计其数的专家的见解。

国际互联网的日益普及正在促进电子邮件作为"一对一"或"一对多"的通信平台。E-mail 允许对等会话，而且如果用户建立了双路通话模式几乎同时就可以收到返回的信息。把文本或图片文件附加在 E-mail 上的功能使用户能把论文和文章发送到任何地址。E-mail 对于教师和学生间的通信极有用，它使学生在学习课程时，遇到的问题能及时得到反馈信息，还可以使教师传送所提交课程的成绩及反馈信息。

　　一般来说，面对面的交流是必要的，而且通过视频电话会议系统能够实现。学生进入互相合作的学习过程，增强了教学体验而减少了单独学习的弊端。双路的全动态视频可通过卫星通信，同轴电缆和在不久的将来要用的光纤连接来发送。目前视频会议的成本限制了它在大多数商业领域的使用，但是随着价格的持续下落它将毫无疑问地用于大多数的远程教育课程。

　　以上讨论的所有技术可以容易地由以计算机为中间媒介的通信来控制，而且许多是计算机技术的扩展。使用以计算机为基础的通信有许多益处。首先是计算机程序具有交互以及反馈信息给学生的能力。其次是计算机能够代替任何已存在的媒体，包括课本和乐器的能力。第三是信息可以以很多种不同的方式表达。更深远的用处包括在模拟模型中使用计算机和设计出能深思的计算机的能力。

2．多媒体与远程教育

　　交互式多媒体资源能够通过多种方式传递给学生。这些方式包含的内容很多，从示证阐述、练习实践和指导（这一部分更多的是教与学中以教师导向为主的方法）到协同合作、解决问题及基于 Web 的教学（这一部分更多的是教与学中以学生导向为主的方法）。

　　在教师导向型的学习方法中，老师是控制学生接收什么样的知识，同时负责向学生传输多少知识的人。在学生导向型的学习方法中，学生能够构筑他们自己的知识体系，并且能将自己的真实经历带到学习过程中来，而这种方式下，教师只被看作是教学负责人。

　　在教师导向型学习方法中，学生在同一个时刻接收的是同样数量的知识，并且在需要阐释和示证这些知识的情况下，学生们能使用的是几种感觉（如视觉、听觉，甚至经验交流）。他们还可以记下并回忆起这些知识点，并且通过练习实践指导这些高度互动式的方法来掌握学科内容。

　　在学生导向型方法里，多媒体素材与互动合作方式结合在一起使用，可以培养团队合作和积极学习的态度，在"问题-解决"方法论的指导下，鼓励更高级别的学习技能，增加学生的理解能力和记忆能力。在学生导向型的指导模型里，学生在学习过程中必须扮演一个积极的角色，必须构筑自己的知识体系或者明确他们学习的目的。学生自己决定如何达到他们预期的学习效果。这种学习方法被嵌入到"建构主义学习理论体系"，这个体系发展于 20 世纪后半叶，并且已具有"认知学习心理学"的基础。在这个建构主义模型中，初学者必须要确立自己学习的目的。在这种情况下，老师不再被认为是唯一权威的教学者，而更可能是一个促进、指导及支持学生自有知识体系的人。

　　如今，多媒体技术作为一种能很好传达指导意见和促进对所交换信息更好的反馈的方法，很多领域的目光都转向了它。多年来，多媒体和多媒体开发者驻守在一些特定的行业里，如广告、娱乐、游戏及计算机辅助训练系统（CBT）。然而，如今的多媒体已经渗透到教育领域，正改变着老师教学和学生学习的方式。随着多媒体技术在教室的应用，教师可以用这些技术来装备自己，变成一个能更好地传授知识的老师，而且可以促进学生以一种更高效的方式进行学习。因此，个人计算机和数字多媒体技术的使用产生了新的学习模式，激发了新颖而富有创造性的学习方法。

此外，随时 20 世纪 90 年代中期互联网和万维网的发展，在教育方法论领域产生了革命性的转变。这场变革将以一个全球化的视野无限扩展我们的学习领域，把学习者和世界范围的教育资源和信息联系在一起。基于 Web 的学习和基于互联网的远程教学使需要它们的人都能够享用。现在，任何一个人都可能通过互联网的连接访问到全世界数不清的图书馆及资讯，把教育行业的面貌改变为信息科技主导的领域。

数字技术的到来是教育行业的福音。同时在最近几年，已经使中国的高等教育机构在他们的教学课程里快速接纳了数字多媒体技术。在我们的教育系统内使用现代科技以加大知识的传授力度，必然反映了这是一个变化的时代。因此，高等教育机构的教育家们如今面临着一个新的挑战，即将这些多媒体技术整合到课堂，改善老师和学生所共有的教学环境。

阅读资料

多媒体技术

多媒体系统必须支持各种媒体类型。这里包括像文本和图形这样平常的媒体，也包括像动画、音频和视频这样丰富的媒体。但是，只有这一点对于多媒体环境是不够的。同样重要的是，要把各种媒体类型的信息源集成到单个系统框架内。因而，一个多媒体系统就是这样一个系统：它允许最终用户以集成的方式共享、通信和处理各种形式的信息。

（1）文本

很多多媒体应用软件是基于由书本到计算机化形式的转换。这种转换使用户能直接访问文本，并为其显示一个弹出式窗口，给出某些具体词的定义。多媒体应用程序也能使用户能够立即显示与正在浏览的某个主题相关的信息。更为强大的是，一本计算机化的书允许用户快速查找信息，而不必根据索引或目录去查找。

（2）声音

把声音融入多媒体程序，用户可以得到使用其他通信方式无法得到的信息。有些类型的信息不用声音是很难有效表达的，比如，用文本文字准确描述心脏跳动以及大海的声音就几乎不太可能。声音也可以加深用户对其他媒体表示的信息的理解。

声音有几种不同的格式。一种是 Windows 的波形文件。波形文件包括用于重放声音的实际的数字数据和信息头，信息头提供有关分辨率及重放速度的附加信息。波形文件可以存储通过麦克风录入的各种声音。另一种可能被采用的声音格式是大家熟知的乐器数字接口，简称 MIDI。MIDI 格式实际是由乐器制造商制定的一种标准，它不是声音的一种数字化格式，而是通过描述要演奏的音符的信息的集合。MIDI 除了音符外不能存储任何文件。

（3）图形图像

因为人类是视觉定位的，因此静态图形图像是多媒体的重要部分。计算机图像处理可定义成对数字表达的图像景象进行的数学函数运算。通常它是视觉感知、模式识别和图像理解所有这些处理的一部分。它们构成了计算机视觉的基础部分。数字图像处理这个术语通常指的是利用数字计算机处理二维画面。从宏观角度上，它概括了任何二维数据的数字处理。一个数字图像是以有限的二进制位数表示的实数或复数数组。以投影胶片、幻灯片、照片或绘图方式给定的图像首先要进行数字化，再以二进制数字阵列形式存储在计算机存储器内。然后，再对这个数字化的图像进行处理，并将其显示在具有高分辨率的电视显示器上。

（4）动画和全运动视频

动画就是运动的图形图像。动画就像静态图形图像一样，都是强有力的信息交流形式。

动画在解释涉及运动的概念时特别有用。全运动视频像电视描绘图像一样，可以使多媒体的应用更加广泛。

7.3 人工智能

课文

人工智能的定义在两个主要方向上有所不同，一个关心的是思维过程和推理，另一个则强调行为。此外，有些定义用人类表现来度量成功，而另一些则用智能的一个理想概念来度量，我们把这个理想概念叫作合理性。如果一个系统能做出正确的事情，那么它就是合理的。所有这些给出了我们在人工智能中要寻求的 4 个可能的目标：像人类那样思维的系统、合理思维的系统、有人类那样行为的系统以及合理行为的系统。

1．人工智能的目标

（1）像人类一样的行为

由阿伦图灵提出的图灵试验是要对智能给出一个满意的、可操作的定义。图灵把智能行为定义为在所有认知工作中能达到人类水平的行为能力，这种能力足以能戏弄询问者。粗略地说，他提出的试验是计算机要由一个人通过电传打字机来向其提出问题，如果这个人不知道另一端是人还是计算机，那么该试验就算通过了。为计算机编写程序让其通过这种试验有许多工作要做，计算机必须具备以下能力：自然语言处理——以使其能用英语（或某种其他人类语言）成功地交互；知识表达——以存储询问之前和询问过程中的信息；自动推理——以利用所存储的信息回答提问并推断出新的结论；以及机器学习——以适应新的环境并发现和推断模式。

（2）像人类一样的思维

如果我们说一个给定的程序能像人一样思维的话，我们必须有一些方法能确定人是怎样思维的。我们需要知道人脑内部的实际工作情况。要做到这一点有两种途径：通过自省——在我们自己的思维流逝的瞬间努力抓住它，或通过心理实验。一旦我们有了一个足够准确的智能理论，就有可能把该理论表示为计算机程序。如果程序的输入/输出和时机选择的行为与人类的行为相当的话，那就证明某些程序的机制对人类也可能有效。认知科学的交叉学科领域把人工智能中的计算机模型和心理学中的实验技术结合在了一起，以试图建立起准确的、可以检测的人类智能活动理论。

（3）合理的思维

19 世纪后期和 20 世纪初期形式逻辑的发展为世界上所有事物及彼此之间关系的陈述提供了一套精确的符号。到 1965 年，出现了这样的程序：如果给定足够的时间和内存，它们就能用逻辑符号来描述一个问题并能求出问题的解（如果解存在的话）。

这种方法有两个主要的障碍。首先，采用非正规知识用逻辑符号所要求的正规术语进行表达是不容易的，尤其是当知识不是 100%确定的时候。其次，在能够用"准则"求解问题和实际求解之间有很大的区别。即使只有几十个论据的问题也能耗尽任何计算机的计算资源，除非它有关于先进行哪一步推理的某种指导。

（4）合理的行为

合理行为是指在给定的信念下能达成目标的行为。代理就是能够意识和行为的某种事物。在这种方法中，人工智能可以被认为是对合理代理的研究和构造。在关于人工智能"思维法则"的方法中，所强调的全是正确性的推理。做出正确的推理有时可以作为一个合理代理的一部分，这是因为一种方法，能够合理地实施也就能够对给定行为达到某人目标的结论进行逻辑推理，然后再按照该结论去行动。另一方面，正确的推理并不是合理性的全部，因为经常有这样的情况存在：在没有证明要做的事情是正确时，有些事情还必须得做。

2．人工智能的应用

（1）语音识别

在 20 世纪 90 年代，计算机语音识别在有限的目标范围内达到了应用水平。因此美国联合航空公司已经用语音识别代替了键盘来记录航班号和城市名这些飞行信息，这非常的便利。另一方面，当指导一些计算机使用语音成为可能时，大多数用户又回归到使用鼠标和键盘，并且也觉得很方便。

（2）理解自然语言

仅仅把一系列的字输入计算机是不够的，分析语句也还是不够的，还必须能给计算机提供一个有助于理解文本涉及领域的知识系统，这个要求在如今非常有限的领域内能够做到。

（3）游戏

你可以买计算机来玩一场价值几百美元的大师级象棋。这其中就有人工智能技术，但是智能技术与人类能够很好地切磋主要是通过极强的计算能力实现的——要分析成百上千个点位。要击败世界冠军就需要这种极强的计算能力，即在每秒内能分析两亿个点位的能力。

（4）专家系统

一个"知识工程师"访问某一领域的专家，并且试图将他们的知识嵌入到计算机程序中以执行一些任务。如何做到这点取决于完成任务所需的智能装置是否处于人工智能现在能达到的水平。当发现并非有这样的智能技术时，就会引起很多令人失望的结果。在 1974 年第一批专家系统中，有一个叫 MYCIN 的系统是用来诊断血液中的细菌性传染病的，并且提供治疗措施。它比医学院的学生和实习医生都做得好，但它的局限性也是明显的，即它本身包括病菌、症状和治疗方法，而不包括病人、医生、医院、死亡、康复和一些突发事件。它只针对一个单独的病人做出反应。因为"知识工程师"通过咨询专家来了解病人、医生、死亡或恢复等信息，很明显，知识专家是利用专家的命令去做预诊框架的。在人工智能的现阶段，这些都是正确的，现在专家系统的有用性取决于用户是否具备常识。

3．机器人学

多数人工智能将最终导致机器人学。多数神经网络、自然语言处理、图像识别、语音识别或语音合成研究工作的最终目标是将它们的技术纳入机器人学的范畴——创造一个完全具有人类特点的机器人。机器人学领域存在的时间几乎与人工智能一样长久。

按照牛津字典的解释，机器人是"外观像人的自动化装置，是智能的、服从的但不受感情影响的机器"。的确，机器人这个词出自于 robota，捷克语的意思是"强迫劳动"。然而，随着机器人学的发展，这个定义很快就变得过时了。从根本上说，机器人被设计成能做人类的工作（除研究机器人学之外），这种工作是乏味的、缓慢的，或者是危险的。直到最近，机器人才开始在工作中使用一定程度的人工智能——许多机器人需要人类操作员，或贯穿在它们的任务中的精确的指示。慢慢地，机器人将会变得越来越独立。机器人和机器之间的区别在于它是否具有自主性、灵活性和精确性。的确，许多机器人只是机器的延伸，但随着该领域的不断进展，它们的区别将会越来越明显。

机器人学正在慢慢走向家庭。最近，机器人学发布了世界上第一个真正的个人机器人——Cye。Cye 允许它的人类操作员建立一张环境图（使用 Windows 界面），并通过工业化机器人链路把它下载到机器人上。于是，该机器人就能漫游该区域，做各式各样的工作——包括用真空吸尘器清扫。

机器人学是一个让许多人感兴趣并绝对令人着迷的领域。随着研究工作逐渐从较为严肃的机器人学项目向商业领域慢慢地渗透，我们将期待更为有趣的（也便宜的）、虚拟宠物的出现。但愿家用商业机器人的价格也不会超过一台价格昂贵的吸尘器。随着计算机变得越来越强大，与你的计算机接口的家用机器人将成为现实，家务活将不再存在。

阅读资料

智能技术

人工智能技术试图了解智能实体，因为智能实体本身很有趣也很有用。很显然具有人类智能水平的计算机将会对我们的日常生活及将来的文明进程产生巨大的影响。我们接下来要进行分析的最常用的人工智能技术是：自然语言、机器人技术、感知系统、专家系统、神经网络和智能软件。

1．自然语言

自然语言包括习惯用语，是指人类使用的语言。自然语言关注计算机语言的识别和语言产生。基本的目标是建立能识别出人类语言并"读出"文字，以及能说和写的计算机的硬件和软件。一个相关目标是制造出这样的软件，它们可以完成人类所要求的研究。

2．机器人技术

机器人技术研究的目标是发展这样的物理系统，它能够完成人类通常所做的工作，特别是在有害或者十分危险的环境中进行的工作。现代机器人技术主要发展数控机床和由计算机辅助制造（CAM）系统驱动的工业制造机器。

3．感知系统

像人类一样，机器人需要"眼睛"和"耳朵"来指导它们的行为。从第二次世界大战以来，计算机科学家和工程师们一直在发展感知系统，或者说是在识别模式意义下，可以"看"和"听"的感觉装置。这个研究领域，有时也被称作"模式识别"，它主要致力于军事应用，比如图片恢复、导弹控制和导航。但这方面的进展并不大，因为教会计算机识别真与假的区别是很困难的。

4．专家系统

专家系统比较接近软件应用，它是在有限的知识和经验领域中寻求专家的意见，并且把它用来解决问题。在人工智能家族中，新闻媒体可能更加关注专家系统。部分原因是当专业支持相当昂贵或短缺时，这样的系统往往能够帮助管理人员和专业人员做出决策。

5．神经网络

人们总是梦想建造可以思考的计算机，一个在某种意义上模仿人类大脑的"电脑"。神经网络就是这种用电流来模仿动物或者人的大脑的生理构造的物理装置。

6．智能软件

（1）模糊逻辑

模糊逻辑是人工智能领域新近兴起的，以规则为基础的计算技术，包括用来再现和推理知识的各种概念和技术，这种再现和推理不准确、不确定或者不可靠。模糊逻辑可以用近似或者主观的数值和不完全的、模糊的数据来建立规则。随着允许使用"高""非常高""极高"这种的表达式，模糊逻辑能够使计算机模拟人们实际中做出决定的方法，而不是用严格的

"IF—THEN"规则来定义问题。

（2）遗传算法

一些人工智能技术用在自然界中发现的方法来解决问题。遗传算法就是一个例子。它们包括多种以达尔文的生物进化论为基础的解决问题的技术。算法开始创建一些模块，这些模块使用像繁殖、变异和自然选择等进程来"繁殖"出结论。当结论变化和合并时，较坏的被抛弃，较好的存活并遗传下去，并且和其他结论一起繁殖出更好的结论。这个过程产生的结果可能会优于其他任何由人类精心计算得到的结果。

（3）智能代理

智能代理的概念在20世纪50年代作为人工智能的一个研究分支得到了发展。从那时起，智能软件代理的影响就越来越大。智能代理为用户完成的任务要求有以下特征：明确性、重复性和可预测性。代理也可以用于商业过程或软件应用，甚至于当它的功能发展到较高水平时，可以很容易超越人的能力。

练习答案

第2章 Hardware Knowledge

2.1 CPU

1. Translate the following phrases into English or Chinese

（1）machine instructions　　　机器指令
（2）binary language　　　二进制语言
（3）arithmetic computations　　　算术运算
（4）current instruction　　　当前指令
（5）instruction register　　　指令寄存器
（6）程序计数器　　　program counter
（7）系统时钟　　　system clock
（8）实时时钟　　　real-time clock
（9）指令周期　　　instruction cycle
（10）微电子技术　　　microelectronic technique

2. Identify the following to be True or False according to the text

（1）T　（2）F　（3）T　（4）F　（5）T　（6）F

3. Reading Comprehension

（1）B　（2）C　（3）D　（4）D　（5）A

2.2 Memory

1. Translate the following phrases into English or Chinese

（1）storage device volume　　　存储设备容量
（2）RAM bars　　　内存条
（3）permanent memory　　　永久性存储器
（4）electrically erasable programmable read-only memory　电可擦除的可编程的只读存储器
（5）expanded memory　　　扩展内存

（6）软件中断　　　　　　　　　software interrupt
（7）存储器芯片　　　　　　　　memory chip
（8）存储器子系统　　　　　　　memory subsystem
（9）二维阵列　　　　　　　　　two dimensional array
（10）输出缓冲区　　　　　　　　output buffer

2．Identify the following to be True or False according to the text
（1）F　（2）T　（3）T　（4）T　（5）F　（6）F

3．Reading Comprehension
（1）A　（2）B　（3）B　（4）D　（5）A

2.3　Input/Output Devices

1．Translate the following phrases into English or Chinese
（1）standard keyboard　　　　　标准键盘
（2）numeric keypad　　　　　　数字键盘
（3）function keys　　　　　　　功能键
（4）capital letters　　　　　　　大写字母
（5）pop-up mean　　　　　　　 弹出式菜单
（6）对话框　　　　　　　　　　dialog box
（7）基于字符的显示器　　　　　character-based display
（8）静电　　　　　　　　　　　static electricity
（9）冷激光打印机　　　　　　　cold laser printer
（10）外置式调制解调器　　　　　external modem

2．Identify the following to be True or False according to the text
（1）T　（2）F　（3）F　（4）T　（5）T　（6）T

3．Reading Comprehension
（1）A　（2）A　（3）B　（4）C　（5）D

第3章　Software Knowledge

3.1　Operating System

1．Translate the following phrases into English or Chinese
（1）interrupt structure　　　　　中断结构
（2）architectural feature　　　　结构特性
（3）context switching　　　　　上下文转接，任务切换
（4）system resource　　　　　　系统资源
（5）static allocation　　　　　　静态分配
（6）应用程序　　　　　　　　　application program
（7）动态分配　　　　　　　　　dynamic allocation
（8）资源分配　　　　　　　　　resource allocation
（9）输入/输出控制系统　　　　　input/output control system

（10）存储保护属性　　　　　　memory protection feature

2. Identify the following to be True or False according to the text

（1）T　（2）F　（3）F　（4）T　（5）F　（6）T

3. Reading Comprehension

（1）B　（2）C　（3）A　（4）D　（5）D

3.2　Data Structures

1. Translate the following phrases into English or Chinese

　　（1）object-oriented software　　面向对象软件
　　（2）algorithmic abstraction　　　算法抽象
　　（3）context-free　　　　　　　　上下文无关
　　（4）stack and queue　　　　　　栈和队列
　　（5）natural language　　　　　　自然语言
　　（6）数据类型　　　　　　　　　data type
　　（7）抽象数据类型　　　　　　　abstract data type
　　（8）原子数据类型　　　　　　　atomic data type
　　（9）结构化数据类型　　　　　　structured data type
　　（10）优先级队列　　　　　　　　priority queue

2. Identify the following to be True or False according to the text

（1）T　（2）T　（3）F　（4）F　（5）T　（6）F

3. Reading Comprehension

（1）C　（2）C　（3）A　（4）A　（5）B

3.3　Programming Language

1. Translate the following phrases into English or Chinese

　　（1）abstract code　　　　　　　抽象代码
　　（2）user-defined type　　　　　用户定义类型
　　（3）machine language　　　　　机器语言
　　（4）data hiding　　　　　　　　数据隐藏
　　（5）grammatical rule　　　　　　语法规则
　　（6）汇编语言　　　　　　　　　assembly language
　　（7）高级语言　　　　　　　　　high-level language
　　（8）面向对象编程　　　　　　　object-oriented programming
　　（9）编程语言　　　　　　　　　programming language
　　（10）机器码　　　　　　　　　　machine code

2. Identify the following to be True or False according to the text

（1）T　（2）T　（3）F　（4）F　（5）T　（6）F

3. Reading Comprehension

（1）C　（2）A　（3）D　（4）C　（5）B

第 4 章 Database Technology

4.1 Database Principle

1. Translate the following phrases into English or Chinese

（1）Database management system　　　　数据库管理系统
（2）Management information system　　　管理信息系统
（3）user-request　　　　　　　　　　　　用户请求
（4）syntax error　　　　　　　　　　　　语法错误
（5）tree structure　　　　　　　　　　　树结构
（6）多用户数据库　　　　　　　　　　　multi-user database
（7）数据处理语言　　　　　　　　　　　data manipulation language
（8）概念模式　　　　　　　　　　　　　conceptual schema
（9）关系结构　　　　　　　　　　　　　relational structure
（10）决策支持系统　　　　　　　　　　decision support system

2. Identify the following to be True or False according to the text

（1）T　（2）T　（3）T　（4）T　（5）F　（6）F

3. Reading Comprehension

（1）B　（2）B　（3）C　（4）D　（5）A

4.2 Data Warehouse and Data Mining

1. Translate the following phrases into English or Chinese

（1）marketing strategy　　　　　　　　　市场策略
（2）sales forecast　　　　　　　　　　　销售预报
（3）enterprise-level data warehouse application　企业级数据仓库应用
（4）data presentation　　　　　　　　　数据呈现
（5）risk management　　　　　　　　　风险管理
（6）决策树　　　　　　　　　　　　　　decision tree
（7）数据仓库　　　　　　　　　　　　　data warehouse
（8）操作数据　　　　　　　　　　　　　operation data
（9）海量存储技术　　　　　　　　　　　mass storage technology
（10）数据挖掘　　　　　　　　　　　　data mining

2. Identify the following to be True or False according to the text

（1）T　（2）T　（3）F　（4）F　（5）T　（6）F

3. Reading Comprehension

（1）B　（2）D　（3）C　（4）A　（5）A

4.3 Big Data and Cloud Computing

1. Translate the following phrases into English or Chinese

（1）Internet-enabled device　　　　　　基于互联网的设备
（2）nontraditional data　　　　　　　　非传统数据

（3）inventory control　　　　　　库存控制
（4）virtual server instance　　　　虚拟服务器实例
（5）enterprise control language　　企业控制语言
（6）云计算　　　　　　　　　　　cloud computing
（7）私有云　　　　　　　　　　　private cloud
（8）公有云　　　　　　　　　　　public cloud
（9）移动电话　　　　　　　　　　mobile phone
（10）大数据　　　　　　　　　　 big data
2．Identify the following to be True or False according to the text
（1）T　（2）T　（3）F　（4）T　（5）F　（6）T
3．Reading Comprehension
（1）A　（2）A　（3）C　（4）A　（5）D

第5章　Computer Network Technology

5.1　Computer Network Basics

1．Translate the following phrases into English or Chinese
　（1）hybrid topology　　　　　　混合网络
　（2）network traffic　　　　　　 网络交通
　（3）next node　　　　　　　　 下一个节点
　（4）campus area network　　　　校园网
　（5）routed network　　　　　　 路由网络
　（6）网络管理员　　　　　　　　network administrator
　（7）局域网　　　　　　　　　　local area network
　（8）网络环境　　　　　　　　　network environment
　（9）物理拓扑　　　　　　　　　physical topology
　（10）逻辑拓扑　　　　　　　　 logical topology
2．Identify the following to be True or False according to the text
（1）T　（2）T　（3）F　（4）F　（5）F　（6）T
3．Reading Comprehension
（1）D　（2）D　（3）C　（4）C　（5）B

5.2　Information Security

1．Translate the following phrases into English or Chinese
　（1）brute force attack　　　　　 暴力攻击
　（2）unauthorized access　　　　 未授权的访问
　（3）cryptographic algorithm　　　加密算法
　（4）attack avoidance　　　　　　攻击避免
　（5）authentication process　　　　验证过程
　（6）访问控制　　　　　　　　　access control

（7）攻击预防　　　　　　　　　　attack prevention
（8）网络安全策略　　　　　　　　network security policy
（9）随机数据流　　　　　　　　　random data stream
（10）安全威胁　　　　　　　　　　security threat

2．Identify the following to be True or False according to the text
（1）T　（2）T　（3）F　（4）F　（5）F　（6）T

3．Reading Comprehension
（1）B　（2）D　（3）B　（4）A　（5）D

5.3　Wireless Networks

1．Translate the following phrases into English or Chinese
（1）access point cell　　　　　　　接入点单元
（2）radio frequency transmission　无线频率传输
（3）infra-red device　　　　　　　远红外设备
（4）service violation　　　　　　　服务侵害
（5）sensitive information　　　　　敏感信息
（6）电磁波　　　　　　　　　　　electromagnetic wave
（7）网络技术　　　　　　　　　　network topology
（8）无线技术　　　　　　　　　　wireless technology
（9）恶意程序　　　　　　　　　　malicious program
（10）无线广域网　　　　　　　　　wireless wide area network

2．Identify the following to be True or False according to the text
（1）F　（2）T　（3）T　（4）F　（5）T　（6）F

3．Reading Comprehension
（1）B　（2）C　（3）D　（4）A　（5）A

第6章　Electronic Commerce

6.1　Electronic Commerce

1．Translate the following phrases into English or Chinese
（1）bank account　　　　　　　　银行账户
（2）digital money　　　　　　　　数字现金
（3）face-to-face　　　　　　　　　面对面
（4）trading partner　　　　　　　贸易伙伴
（5）geographical distance　　　　地理距离
（6）电子数据交换　　　　　　　　electronic data interchange
（7）信用卡　　　　　　　　　　　credit card
（8）电子媒介　　　　　　　　　　electronic media
（9）电子支付系统　　　　　　　　electronic payment system
（10）集成电路　　　　　　　　　　integrated circuit

2. Identify the following to be True or False according to the text
（1）F　（2）F　（3）F　（4）T　（5）T　（6）T
3. Reading Comprehension
（1）B　（2）B　（3）A　（4）D　（5）C

6.2　Web Navigation

1. Translate the following phrases into English or Chinese
（1）web navigation　　　　　网站导航
（2）page layout　　　　　　页面布局
（3）design presentation　　　设计展示
（4）top-level page　　　　　顶层页面
（5）navigational link　　　　导航链接
（6）信息组织　　　　　　　information organization
（7）两阶段扫描　　　　　　two-stage scanning
（8）导航系统　　　　　　　navigation system
（9）导航设计　　　　　　　navigation design
（10）动态菜单　　　　　　　dynamic menu
2. Identify the following to be True or False according to the text
（1）T　（2）T　（3）T　（4）F　（5）F　（6）T
3. Reading Comprehension
（1）D　（2）A　（3）A　（4）B　（5）C

6.3　Mobile Commerce

1. Translate the following phrases into English or Chinese
（1）high rate of growth　　　高增长率
（2）mobile marking　　　　　移动营销
（3）content provider　　　　内容提供商
（4）mobile inventory　　　　移动库存
（5）mobile station　　　　　移动基站
（6）桌面电子商务　　　　　desktop e-commerce
（7）银行账户　　　　　　　bank account
（8）移动商务　　　　　　　mobile commerce
（9）零售物品　　　　　　　retail goods
（10）智能卡　　　　　　　　smart card
2. Identify the following to be True or False according to the text
（1）T　（2）F　（3）T　（4）T　（5）F　（6）T
3. Reading Comprehension
（1）A　（2）B　（3）C　（4）B　（5）D

6.4 Electronic Payment and Logistics

1．Translate the following phrases into English or Chinese
（1）electronic check   电子支票
（2）stock quotation   股票行情
（3）digital cash software   数字现金软件
（4）fund transfer   资金转移
（5）logistics manager   物流经理
（6）电子钱包   electronic wallet
（7）数字签名   digital signature
（8）物流信息系统   logistics information system
（9）订单处理系统   order processing system
（10）库存状态   inventory status

2．Identify the following to be True or False according to the text
（1）T （2）T （3）T （4）T （5）F （6）F

3．Reading Comprehension
（1）C （2）A （3）D （4）B （5）B

第7章 Computer Applications

7.1 Office Automation

1．Translate the following phrases into English or Chinese
（1）voice recognition and synthesis   语音识别和合成
（2）statistic chart   统计图
（3）satellite broadcasting   卫星广播
（4）desktop publishing   桌面印刷
（5）integrated information processing network   集成信息处理网络
（6）键盘布局   keyboard layouts
（7）数据处理   data processing
（8）办公自动化   office automation
（9）字处理程序   word processing program
（10）电子表格程序   spreadsheet program

2．Identify the following to be True or False according to the text
（1）F （2）F （3）T （4）T （5）T （6）F

3．Reading Comprehension
（1）D （2）C （3）A （4）A （5）C

7.2 Distance Education

1．Translate the following phrases into English or Chinese
（1）distance education   远程教育
（2）information storage   信息存储

（3）optical technology　　　　光学技术
（4）news service　　　　　　　新闻服务
（5）educational tools　　　　　教育工具
（6）电子公告板　　　　　　　　electronic bulletin board
（7）电子通信　　　　　　　　　electronic communication
（8）仿真模型　　　　　　　　　simulation model
（9）通信平台　　　　　　　　　communication platform
（10）点对点　　　　　　　　　 peer-to-peer

2. Identify the following to be True or False according to the text
（1）F　（2）F　（3）T　（4）T　（5）T　（6）T

3. Reading Comprehension
（1）A　（2）A　（3）B　（4）C　（5）D

7.3　Artificial Intelligence

1. Translate the following phrases into English or Chinese
（1）human performance　　　　人类行为
（2）thought process　　　　　　思维过程
（3）image recognition　　　　　图像识别
（4）cognitive science　　　　　 认知科学
（5）logical notation　　　　　　逻辑标记
（6）神经网络　　　　　　　　　neural networking
（7）语音识别　　　　　　　　　speech recognition
（8）人工智能　　　　　　　　　artificial intelligence
（9）专家系统　　　　　　　　　expert system
（10）虚拟宠物　　　　　　　　 virtual pets

2. Identify the following to be True or False according to the text
（1）F　（2）T　（3）T　（4）T　（5）T　（6）F

3. Reading Comprehension
（1）D　（2）C　（3）B　（4）A　（5）D

附录 1 计算机专业英语词汇表

A

abnormal end 异常终止
abstract data type 抽象数据类型
acceleration card 加速卡
access control 访问控制
accessibility 易接近的，可到达的
access list 访问控制表
access permission 访问许可
access time 存取（访问）时间
accessory program 附件程序
account 账号
accounting software 会计软件
acoustic 有关声音的，声学的
acronym 缩略词
active desktop 活动桌面
active window 活动窗口
acyclic directory structure 非循环目录结构
adapter card 适配卡
adaptive scheduler 自适应调度
addressing mechanism 寻址机制
address space 地址空间
administrator 管理员
algorithm 算法，规则系统
alignment 队列
alphabetic 依字母顺序的
alphanumeric 文字数字的

amplify 放大
animation 动画
anti-virus program 防病毒程序
application integration 应用程序集成
application layer 应用层
application object 应用对象
archiving 存档
arrow keys 箭头键，方向键
artificial intelligence 人工智能
assembler 汇编程序
assessment 评估
assignment 分配
assortment 分类
asymmetric encryption 非对称加密
asynchronous 异步的
asynchronous primitive 异步原语
atomic action 原子操作
atomicity property 原子属性
attachment 附件
attribute 属性，标志
auction online 在线拍卖
audience demographics 受众人数
authentication 鉴别，证实
authorization 授权，认可
automation server 自动化服务器
auxiliary 辅助的

B

background 后台，背景
bandwidth 带宽
Banner 旗帜广告
bar code 条形码
base-band 基带
batch processing 成批处理
baud 波特
baud rate 波特率
bibliography 参考书目
big data 大数据
binary digit 二进制数字
bind 绑定
biometrics 生物统计学

biometrics device　生物特征辨识装置
bitable　双稳态的
bitmap　位图
black box　黑盒子
block diagram　框图
block structure　模块化结构
Boolean logic　布尔逻辑
boot block　引导块
boot sector　引导扇区
boot strap　引导程序
boot up　启动
boundary　界限
branch　分支
broadband network　宽带网络
broadcast address　广播地址
broadcast storm　广播风暴
browser　浏览器
bubble jet printer　喷墨打印机
built-in　内装的，固有的
bulk storage　大容量存储器
bulletin board　告示板，公告板
bus-contention　总线争用

C

calibration　校准，定标
carriage return　回车
carrier　载波
cartography　绘图法
certificate authority　证书认证
chain reaction　链式反应
channel　信道
chassis　机箱
check box　复选框
child node　子节点
child window　子窗口
chipset　芯片组
cipher text　密文
circuit switching　电路交换
classification　分级，分类
client program　客户程序
clipboard　剪贴板

client/server 客户机/服务器
cloud computing 云计算
coaxial cable 同轴电缆
coding 编码
cognitive 认知的，认识的
coherent 一致的，连贯的
collaborative 合作的，协作的
command button 命令按钮
comment 注释
communication deadlock 通信死锁
compatibility 兼容性
compiler 编译程序，编译器
compression 压缩
computability 可计算性
computerize 计算机化
concurrent 并发的，同时发生的事件
confidential 秘密的，机密的
configuration 配置，构造
conflict 冲突
congestion 拥塞
connectionless service 无连接服务
connection-oriented service 面向连接的服务
constraint 约束，强制
construct 构造，创立
container 容器，一种特殊的屏幕区域和组件，其中可包含组件
context 上下文
control box 控制框
control panel 控制面板
cooling fan 冷却风扇
coordinate 协调
cordless mouse 无绳鼠标
core 磁芯
correlation 相互关系，相关性
cracker 解密者
criteria 标准
critical region 临界区
cross platform 跨平台的
cryptography 密码学
cybercash 电子货币
cybercrime 网络犯罪

cyberspace 电脑空间

D

data flow diagram 数据流程图
data link layer 数据链路层
data mining 数据挖掘
data model 数据模式
data source 数据源
data stream 数据流
data structure 数据结构
data transfer rate 数据传输速度
data warehouse 数据仓库
data window object 数据窗口对象
database 数据库
database engine 数据库引擎
database interface 数据库接口
database server 数据库服务器
deadlock 死锁
deadline 最终期限
debit-card 借记卡
debugger 调试程序
decision tree 决策树
decoder 译码器
decryption 解密
definition 定义
deformation 变形
digital signature 数字签名
digital wallet 数字钱包
demodulator 解调器
demographic 人口统计的
desktop 桌面
destination 目的地，目标文件
device contention 设备竞争
device dependent 设备相关的
device independent 设备无关的
device object 设备对象
diagnosis 诊断
diagram 图表
dialog box 对话框
digital camera 数码相机
digital cash 数字现金

digital certificate　数字证书
digital signature　数字签名
directory　目录
discrete mathematics　离散数学
disk cleanup　磁盘清理
disk defragmenter　磁盘碎片整理工具
disk drive　磁盘驱动器
diskette　磁盘
diskless workstation　无盘工作站
display adapter　显示适配器
dissertation　论文，专题
distributed database　分布式数据库
distributed processing　分布式处理
distributed system　分布式系统
document　文档，文件，资料
dot-matrix　点阵
downstream　向下传输
drop-down listbox　下拉式列表框
drop-down menu　下拉式菜单
drum　硒鼓
dynamic binding　动态绑定
dynamic IP address　动态 IP 地址
dynamic page　动态页面

E

eavesdropping　窃听
edit　编辑
electronic cash　电子现金
electronic check　电子支票
electronic commerce（E-commerce）电子商务
electronic wallet　电子钱包
electronic mail（E-mail）电子邮件
electronic meetings　电子会议
electronic wallet　电子钱包
efficiency　有效性，效率
electronic banking　电子银行
electronic money　电子货币
electronic payments　电子支付
embedded computer　嵌入式计算机
embedded real-time system　嵌入式实时系统
emoticons　表情符号

emulation 仿真
encapsulation 封装，将相关的数据和过程打包在一个对象中
encode 编码
encryption 加密
encryption key 加密密钥
end user 终端用户
entity 实体，OSI 模型中活跃在每一层的单元
enquiry 询问
enumerate 枚举，列举
equivalent 相等的，相当的
erasure 删去，消除
etched circuit 蚀刻电路
etching technology 蚀刻技术
Ethernet 以太网
evolution 评价，估计
exception 异常
expanded memory 扩充内存
expansion slot 扩展插槽
expertise 专业知识
expert system 专家系统
exponentially 指数的，幂的
extended attributes 扩展属性
extended memory 扩展内存
extension 扩展名

F

facilitate 使方便，促使
facility 功能，工具
factor 因素
fatal error 致命错误
fault tolerance 容错
feasibility 可行性
feature 特征，特色
feedback 反馈
fiber-optic cable 光纤
field 字段，数据库中表的每一列称为一个字段
file handle 文件句柄
file server 文件服务器
file system 文件系统
filter 过滤器
fingerprint scanner 指纹扫描仪

firewall   防火墙
firmware   固件
flash memory   闪存
flexibility   弹性，适应性
floppy disk   软盘
flowchart   流程图，框图
flow control   流量控制
folder   文件夹
font format   字样格式
foreground job   前台作业
foreign agent   外地代理
foreign key   外键，数据库中用以建立同其他表间的关联
format   格式化
fragmentation   （程序的）分段存储，存储（碎）片
frame   帧
front-end   前端，前台
function key   功能键

G

game theory   博弈论
gateway   网关
gigabit network   千兆网
global scheduler   全局调度
Gopher   一个著名的文档检索工具
graphics package   图形软件（包）
graphics tablet   图形输入板
grayscale   灰色标度
grid   格子，栅格
group editor   群编辑器
groupware   组件
guarantee   保证
guidance   向导，指导

H

hacker   黑客
hanging indent   悬挂式缩进
hardcopy   硬拷贝
hashed file   散列文件
headline   大字标题
head pointer   头指针
head node   头节点
header and footer   页眉和页脚

heap sort 堆排序
hexadecimal system 十六进制
hierarchical directory structure 层次目录结构
high-level language 高级语言
histogram 柱状图
home page 主页
host computer 宿主机
hub 集线器
Huffman codes 赫夫曼编码
Huffman tree 赫夫曼树
hyperlink 超链接
hypermedia 超媒体
hypertext 超文本
hypothetical 假设的，假定的

I

icon 图标
image 图像
image map 图像映射
immoveable 固定的，不可移动的
implementation 执行
index 索引
individual 个人，个体
infection 传染
information superhighway 信息高速公路
inheritance 继承（面向对象的三大特性之一）
initialized 已初始化的
instruction 指令
integrated package 集成软件包
integration 综合
integrity 完整性
intelligent bridge 智能网桥
intercept 截获
interception 侦听
interface 接口，界面
interlacing 隔行扫描
internal memory 内存储器
Internet 互联网
Internet of things 物联网
Internet telephone 网络电话
interpreter 解释器，解释程序

interrupt 中断
interval 时间间隔
int-jet printer 喷墨打印机
Intranet 企业内部互联网
inventory 详细目录，库存货
inversion 反相，倒置
IP address 网际协议地址
iterative process 迭代过程

J

job object 作业对象
job scheduler 作业调度程序
joystick 操纵杆
junk E-mail 垃圾邮件
jurisdiction 权限
justification 对齐，版面调整
just-in-time manufacturing 即时生产

K

kernel 核心
keyboard 键盘
keyguard 键盘守卫
keypad 辅助小键盘
keyword 关键字
kilobit 千比特
kilobyte 千字节
kit 用具包

L

label 标记，记号
laser-etched 激光蚀刻的
laser printer 激光打印机
layer 层
layout 规划，布局
leased line 专线
legitimacy 合法性，正统性
letter quality 字符模式
library 库
life cycle 生命周期
light-pen 光笔
linear linked lists 线性链表
linear lists 线性表
link 链接

linkage 链接
linked radix sort 链式基数排序
load 装载，装入
local scheduler 本地调度
local variable 局部变量
location 定位，位置，配置
logical fashion 逻辑方式
logical link control 逻辑链路更新
logical schema 逻辑模式
logic circuit 逻辑电路
logic complementation 逻辑补码法
login 注册
logistics 后勤，物流
log on 登录
log out 注销登录
low-level language 低级语言

M

machine code 机器码
machine language 机器语言
macro 宏
magnetic pot 磁场
magnetic tape 磁带
magnification 扩大，放大倍率
mailbomb 邮件炸弹
mail merging 邮件合并
mainframe 大型机，主机
main memory 主存储器
maintain 维护
maintenance 维持，维护
main window 主窗口
malignant 恶性的
manipulation 处理，操作
manufacturer 制造者，厂商
masked edit 屏蔽编辑
match 匹配，相配
matrix 矩阵
mechanism 机构，机制
media player 媒体播放器
medium 媒介，媒体
megabit 兆位，百万位

megabyte 兆字节
megahertz 兆赫兹
memory card 内存卡
memory stick 内存棒
menu bar 菜单栏
metadata 元数据
microelectronic 微电子的
microphone 麦克风
microwave 微波
middleware 中间件
miniaturize 使小型化
minicomputer 小型计算机
mirror 镜像
mobile commerce 移动商务
mobile marking 移动营销
mobile phone 移动电话
Modem 调制解调器
monochrome 单色的，黑白的
monopoly 垄断，专利权
mother board 系统板
mount 装配
multicomputer 多计算机系统
multidocument interface 多文档界面
multiline edit box 多行编辑框
multimedia 多媒体
multiple inheritance 多重继承
multi-threaded 多线程
multi-processor 多处理器
multitasking 多任务
mutual exclusion 互斥

N

nanosecond 纳秒
natural language 自然语言
navigate 导航
netiquette 网络礼仪
network layer 网络层
network system 网络系统
network administer 网络管理员
neural networking 神经网络
newsgroups 新闻讨论组

niche 小生境
non-blocking primitive 非阻塞原语
non-impact 非击打式
nonrepudiationg 不可抵赖性
normalization 标准化，正常化
notation 表示法，符号，标志
notepad 记事本
nozzle 喷头
null string 空串
numerical 数字的
Num Lock 数字键锁定

O

object-based system 基于对象的系统
object-oriented 面向对象
object-oriented system 面向对象系统
octal system 八进制系统
offline 离线
online 在线
open system 开放系统
operand 操作数
operational 操作的
optical disk 光盘
optical fiber cable 光导纤维电缆
optimal scheduling algorithm 最优调度算法
optimal tree 最优树
optimization 最优化，最佳化
ordered tree 有序树
orthogonal list 十字链表
outline view 大纲视图
overflow 上溢
overfrequency 超频
overlapped 重叠
overloading 重载
out-dated 过期的，逾期的
output device 输出设备
outsourcing 外包

P

package 软件包，组件
page description language 页面描述语言
page fault 页故障

parallel port　并行接口
painterbar　一个按钮栏，允许访问一个给定描绘器中的所有主要功能
paradigm　范例
password　口令，密码
perform　执行
peripheral　外围设备
permission　许可，特制资源拥有者对共享该资源的人的信任程度
photo-sensitive drum　感光鼓
pie chart　饼图
pin printer　针式打印机
pirate　盗版者，盗版
plaintext　明文
platform　平台
plotter　绘图仪
plug and play（or PnP, P&P）　即插即用，是用于解决设备安装麻烦的一套技术方案
pointing device　定位设备
polling task　轮流查询任务
polymorphism　多态，同一个对象中的两个或多个名字相同、参数列表不同的函数
populate　板上组装
pop-up menu　弹出式菜单
pop-up Window　弹出式窗口
portability　可移植，轻便
portable　便携的
postulate　假定，基本条件
precision transforms　精确度变换
preemptive multitasking　抢先式多任务
printed circuit board　印刷电路板
printer driver　打印机驱动程序
Primary Key　主键，唯一标识数据库表中每条记录的一个或多个列
private cloud　私有云
private key cryptography　私钥加密
privileged instruction　特权指令
procedural programming　面向过程程序设计
process　进程
protocol　协议
prototype　原型
proxy server　代理服务器
public cloud　公有云
public key　公开密钥
public key cryptography　公钥加密

pull technology   拉式技术
push technology   推式技术

Q

quad speed   四倍速
quadratic probing   二次探测
quantizer   数字转换器，编码器
quantometer   光谱分析仪
query   查询
queue   队列
quit   退出

R

radio button   单选按钮
rationale   基本原理
readiness   准备就绪
real time system   实时系统
recipient   收件人
recorder   记录器
recursive function   递归函数
refresh   刷新
refresh time   刷新率，更新率
relational model   关系模型
remote terminal   远程终端
replicate   重复
remark   注释
remote   远程
removable   可移动的
replicate   复制
reproduction   复制品
reset   复位
resident   驻留的
resolution   分辨率
response window   响应式窗口
restore   恢复
retrieve   检索
revision   修改，修正
ribbon cartridge   色带盒
right-click   右击
ring network   环形网络
robotics   机器人技术
router   路由器

routine 日常工作，例程
S
safe mode 安全模式
scalability 可伸展性
scale 定标，缩放
scanner 扫描仪
screen saver 屏幕保护程序
script 脚本
seamless 无缝连接的
search engine 搜索引擎
sector 扇区
security 安全
segment 段，节
serial port 串行接口
sector 扇区
security certificate 安全认证
sensor 传感器
sequential 连续的
server 服务器
session 会话
shared variable 共享变量
short cut 短路
shortcut key 快捷键
signature 签名
simulator 仿真器，模拟器
simultaneous 同时发生的
solid ink 固体油墨
sound box 音箱
sound card 声卡
source code 源代码
span 跨越
spam 垃圾邮件
spatiality 空间性
speech generator 译音发生器
speech recognition 语音识别
speech synthesizer 语音合成器
spreadsheet 电子表格
spyware 间谍软件
stack 堆栈
stereo 立体音响，立体感觉的

storage class specifier 存储类标识符
streaming audio 音频流
streaming video 视频流
streamline 流线型的
structure 结构体
structure chart 结构图
subnet 子网
subnet mask 子网掩码
subroutine 子程序
supercomputer 超级计算机
supplier 供应者，厂商
switch 交换机
symbolize 象征，用符号表现
syntax error 语法错误
synthesizer 综合者，合成器
synchronous 同步

T

taskbar 任务栏
technologist 技术专家
technology 工艺，科技
telecommunicating 电子通信
telemarketing 电话销售
telemetry 遥感勘测
telnet 远程登录程序
template file 模板文件
temporality 暂时，此时
tender 投标
terminal 终端
terminology 术语学
thermal printer 热敏打印机
thin client 瘦客户端
thread 线程
three-dimensions 三维
throughput 吞吐量
thumbnail 缩略图
time-sharing 分时
time slicing 时间分片
time-varying 时间变换的
title bar 标题条
toggle switch 拨动开关

token 令牌，记号
toolkit 工具包
topology 拓扑
touchpad 触摸板
touch-sensitive display 触控式显示器
trackball 轨迹球
transaction 事务
transceiver 收发器
translator 翻译程序
transparency 透明，透明度
transport layer 传输层
Turing Test 图灵试验
typeface 字体，字样
typeset 排版
typewrite 打字机

U

ubiquitous computing 普适计算
unambiguous 不含糊的，清楚的
unauthorized access 未授权访问
undirected graph 无向图
union 共同体
uni-programming 单道程序设计
unleash 释放
unordered tree 无序树
update 更新，修正
upgrade 升级
upload 上传
upstream rate 上行速率
user account 用户账号
user-defined 用户自定义
user ID 用户标识符
User Object 用户对象
utility 实用程序
utilize 利用

V

vacuum tube 真空管
valid 有效值
variable 变量
vector 矢量
vendor 供应商

versatile   通用的，万能的
version   版本
vertical   垂直的
video bandwidth   视频带宽
video capture card   视频采集卡
video clips   视频片段
video conferencing   电视会议
video display   视频显示
video phone   可视电话
virtual address space   虚拟地址空间
virtual block caching   虚拟块高速缓存
virtual device   虚拟设备
Virtual Host Service   虚拟主机服务
virtual interface   虚拟接口
virtual IP address   虚拟 IP 地址
virtual memory technology   虚拟存储器技术
virus checker   病毒检查程序
visual arts   视觉艺术
visualize   形象，形象化
voice mail   语音邮件
voice synthesis   语音合成
volatile   易失性的
volume label   卷标
voice control   语音控制

W

warm boot   热启动
wave form   波形
wave length   波长
Web page   网页
Web paging   网页寻呼
Web server   Web 服务器
Web site   Web 站点
wheel   特权用户
wholesaler   批发商
Wild Card Character   通配符
Window-based   基于视窗的
Window Painter   窗口画板
Windows   窗口
Windows message   窗口消息
wire pirate   网盗，企图在网络中窃取口令并且访问未授权的系统和数据的人

wiretapping 搭线窃听
wizard 向导工具
word processor 文字处理软件
workgroup hub 工作组集线器
worksheet 工作表
workspace 工作区
workstation 工作站
worm 蠕虫
worthless 无价值的，无益的
wrap 隐藏，伪装
write protect 写保护

X

xerography 干印法，静电复印术

Y

yield 出产
yoke 磁头组

Z

zero access 立即存取
zero complement 补码
zero suppression 消零
zone 区域
zoom 放大

附录 2 计算机专业英语缩写词表

A

AAC（Activity Address Code） 有效地址代码
AAL（ATM Adaptation Layer） ATM 适配层
AAR（Automatic Address Recognition） 自动地址识别
AAS（Automatic Audio Switch） 自动音频变换技术
AAT（Average Access Time） 平均存取时间
AB（Address Bus） 地址总线
ABEOJ（Abnormal End of Job） 作业异常终止
ACL（Access Control Lists） 访问控制表
ACK（Acknowledgement Character） 确认字符
ACM（Association for Computer Machinery） 计算机协会
ADC（Analogue to Digital Converter） 模数转换器
ADSL（Asymmetric Digital Subscriber Line） 非对称用户数字线路
AGP（Accelerated Graphics Port） 图形加速端口
AIFF（Audio Image File Format） 声音图像文件格式
ALU（Arithmetic Logic Unit） 算术逻辑单元
ANSI（American National Standard Institute） 美国国家标准协会
API（Application Programming Interface） 应用程序设计接口
APPN（Advanced Peer-to-Peer Network） 高级对等网络
ARP（Address Resolution Protocol） 地址分辨/转换协议
ASCII（American Standard Code for Information Interchange） 美国信息交换标准代码
ASP（Application Service Provider） 应用服务提供商
AT（Asynchronous Transmission） 异步传输
AT&T（American Telephone and Telegraph Company） 美国电报电话公司
ATM（Asynchronous Transfer Mode） 异步传输模式
ATM（Automatic Teller Machine） 自动柜员机

AVI（Audio Video Interface） 声音视频接口
AWC（Active Wire Concentrator） 集线器

B

B2B（Business to Business） 商业机构对商业机构的电子商务
B2C（Business to Consumer） 商业机构对消费者的电子商务
BBA（Broad Band Access） 宽带接入
BBS（Bulletin Board System） 电子公告牌系统
BC（Bar Code） 条形码
BCC（Basic Connection Components） 基本连接组件
BCE（Buffer Control Element） 缓冲器控制单元
BCIU（Bus Control Interface Unit） 总线控制接口单元
BCLK（Bus Clock） 总线时钟
BD（Bus Driver） 总线驱动器
BDR（Bus Device Request） 总线设备请求
BGP（Border Gateway Protocol） 边缘网关协议
BIOS（Basic Input/Output System） 基本输入/输出系统
BISDN（Broadband-Integrated Services Digital Network） 宽带综合业务数字网
BLU（Basic Link Unit） 基本链路单元
BOF（Beginning Of File） 文件开头
BPS（Bits Per Second） 每秒比特数
BRI（Basic Rate Interface） 基本速率接口
BSC（Bus System Control） 总线系统控制
BSP（Byte Stream Protocol） 字节流协议
BSS（Broadband Switching System） 宽带交换系统

C

CA（Certificate Authority） 证书认证
CAD（Computer Aided Design） 计算机辅助设计
CAE（Computer-Aided Engineering） 计算机辅助工程
CAI（Computer Aided Instruction） 计算机辅助教学
CAM（Computer Aided Manufacturing） 计算机辅助管理
CAS（Control Automatic System） 自动化控制系统
CASE（Computer Assisted Software Engineering） 计算机辅助软件工程
CAT（Computer Aided Test） 计算机辅助测试
CATV（Community Antenna Television） 有线电视
CB（Control Bus） 控制总线
CCS（Common Channel Signaling） 公共信令
CDFS（Compact Disk File System） 密集磁盘文件系统
CD-DA（Compact Disc-Digital Audio） 数字音乐光盘
CDMA（Code Division Multiple Access） 码分多址技术
CD-MO（Compact Disc-Magneto Optical） 磁光式光盘

CD-ROM（Compact Disc Read-Only Memory） 只读光盘
CGI（Common Gateway Interface） 公共网关接口
CMS（Color Management System） 色彩管理系统
COM（Component Object Model） 组件对象模型
CORBA（Common Object Request Broker Architecture） 公共对象请求代理结构
CPU（Central Processing Unit） 中央处理单元
CR（Carriage Return） 回车符
CRC（Cyclical Redundancy Check） 循环冗余校验码
CRM（Client Relation Management） 客户关系管理
CRT（Cathode-Ray Tube） 阴极射线管，显示器
CSS（Cascade Style Sheets） 层叠样式表
CTS（Clear To Send） 清除发送
CU（Control Unit） 控制单元

D

DA（Data Administrators） 数据管理者
DAC（Digital to Analogue Converter） 数模转换器
DAE（Digital audio Extraction） 数字音乐析取
DAO（Data Access Object） 数据访问对象
DAT（Digital Audio Tape） 数字式音频磁带
DAT（Disc Allocation Table） 磁盘分配表
DB（Data Bus） 数据总线
DBA（Data Base Administrator） 数据库管理员
DBCS（Data Base Control System） 数据库控制系统
DAP（Directory Access Protocol） 目录访问协议
DBMS（Data Base Management System） 数据库管理系统
DCE（Data Communication Equipment） 数据通信设备
DCE（Distributed Computing Environment） 分布式计算环境
DCOM（Distributed COM） 分布式组件对象模型
DD（Data Dictionary） 数据字典
DDB（Distributed Data Base） 分布式数据库
DDE（Dynamic Data Exchange） 动态数据交换
DDI（Device Driver Interface） 设备驱动程序接口
DDK（Driver Development Kit） 驱动程序开发工具包
DDN（Distributed Data Network） 分布式数据网
DEC（Digital Equipment Corporation） 数字设备公司
DES（Data Encryption Standard） 数据加密标准
DHCP（Dynamic Host Configuration Protocol） 动态主机配置协议
DLL（Dynamic Link Library） 动态链接库
DM（Data Mining） 数据挖掘
DMA（Direct Memory Access） 直接内存访问

DMSP（Distributed Mail System Protocol） 分布式电子邮件系统协议
DNA（Distributed Network Architecture） 分布式网络结构
DNS（Domain Name System） 域名系统
DNS（Domain Name Server） 域名服务器
DOM（Document Object Mode） 文档对象模型
DOS（Disk Operation System） 磁盘操作系统
DRAW（Direct Read After Write） 写后直接读出
DSM（Distributed Shared Memory） 分布式共享内存
DSP（Digital Signal Processing） 数字信号处理
DSS（Decision Support System） 决策支持系统
DT（Data Terminal） 数据终端
DTD（Document Type Definition） 文件定义类型
DTE（Data Terminal Equipment） 数据终端设备
DTV（Digital Television） 数字电视
DVC（Digital Video Camera） 数码摄像机
DVD（Digital Versatile Disc） 数字多功能盘
DVI（Digital Video Interactive） 数字视频交互

E

EAR（Effective Address Register） 有效地址寄存器
ECI（Electronic Customer Interchange） 电子客户交换机
ECP（Error Control Procedure） 差错控制过程
EDI（Electronic Data Interchange） 电子数据交换
EDIF（Electronic Data Interchange Format） 电子数据交换格式
EDP（Electronic Data Processing） 电子数据处理
EEPROM（Erasable and Electrically Programmable ROM） 电擦除可编程只读存储器
EFT（Electronic Funds Transfer） 电子汇款，电子资金转账
EGP（External Gateway Protocol） 外部网关协议
EISA（Extended Industry Standard Architecture） 增强工业标准结构
EMM（Expanded Memory Manager） 扩充内存管理程序
EMS（Expanded Memory Specification） 扩充存储器规范
EMS（Electronic Mail Service） 电子邮件业务
EPH（Electronic Payment Handler） 电子支付处理系统
EPROM（Erasable Programmable ROM） 可擦除可编程只读存储器
ERP（Enterprise Resource Planning） 企业资源计划
ERTS（Earth Resources Technology Satellite） 地球资源技术卫星

F

FACS（Fully Automatic Compiling System） 全自动编译系统
FAN（Fiber Access Network） 光纤接入网
FAQ（Frequently Asked Questions） 常见问题解答
FAT（File Allocation Table） 文件分配表

FCB（File Control Block） 文件控制块
FCFS（First Come First Service） 先到先服务
FCP（Firewall Control Protocol） 防火墙控制协议
FCS（Frame Check Sequence） 帧校验序列
FDD（Floppy Disk Device） 软盘驱动器
FDDI（Fiber-optic Data Distribution Interface） 光纤数据分布接口
FDM（Frequency-Division Multiplexing） 频分多路
FDMA（Frequency Division Multiple Address） 频分多址
FDX（Full DupleX） 全双工
FEK（File Encryption Key） 文件密钥
FIFO（First In First Out） 先进先出
FMP（File Management Program） 文件管理程序
FNN（Fuzzy Neural Network） 模糊神经网络
FPU（Floating-Point Unit） 浮点部件
FRC（Frame Rate Control） 帧频控制
FTP（File Transfer Protocol） 文件传输协议
FTR（File Transfer Request） 文件传送请求
FTT（Fault Tolerance Technology） 容错技术

G

GAI（Graphics Adapter Interface） 图形适配器接口
GAL（General Array Logic） 通用逻辑阵列
GAP（Gateway Access Protocol） 网关存取协议
GB（GigaByte） 千兆字节
GCS（Ground Communication System） 地面通信系统
GCR（Group-Coded Recording） 成组编码记录
GDI（Graphics Device Interface） 图形设备接口
GDP（Graphic Data Processing） 图形数据处理
GIS（Geographic Information System） 地理信息系统
GPI（Graphical Programming Interface） 图形编程接口
GPIB（General Purpose Interface Bus） 通用接口总线
GPMS（General Personnel Management System） 通用人事管理系统
GPS（Global Positioning System） 全球定位系统
GPU（Graphics Processing Unit） 图形处理器
GSM（Group Special Mobile） 分组专用移动通信
GSX（Graphics System Extension） 图形系统扩展
GUI（Graphical User Interface） 图形用户接口
GVPN（Global Virtual Private Network） 全球虚拟专用网

H

HAMT（Human-Aided Machine Translation） 人工辅助机器翻译
HCI（Human Computer Interface） 人机接口

HDC（Hard Disk Control） 硬盘控制器
HDD（Hard Disk Drive） 硬盘驱动器
HDDR（High Definition Digital Recorder） 高清晰度数字录像机
HDLC（High-level Data Link Control） 高级数据链路控制
HDTV（High-Defination Television） 高清晰度电视
HDX（Half DupleX） 半双工
HFS（Hierarchical File System） 分层文件系统
HIM（Hardware Interface Module） 硬件接口组件
HK（Hot Key） 热键
HPNA（Home Phoneline Network Alliance） 家用电话线网络联盟
HSDL（High-Speed Data Link） 高速数据链路
HTML（Hyper Text Markup Language） 超文本标记语言
HTTP（Hyper Text Transport Protocol） 超文本传输协议
HYDAC（HYbrid Digital Analog Computer） 混合式数字模拟计算机

I

IAC（Inter-Application Communications） 应用间通信
IBM（International Business Machines）美国国际商用机器公司
ICC（Integrated Circuit Card） 集成电路卡
ICD（Interface Control Document） 接口控制文档
ICMP（Internet Control Message Protocol） 因特网控制消息协议
ICP（Internet Content Provider） 因特网内容服务提供商
ICN（International Computer Network） 集成计算机网络
IDC（International Development Center） 国际开发中心
IDE（Integrated Development Environment） 集成开发环境
IDL（Interface Description Language） 接口描述语言
IDL（Interface Definition Language） 接口定义语言
IDN（Integrated Data Network） 综合数据网
IDS（Intelligence Data System） 智能数据系统
IDU（Interface Data Unit） 接口数据单元
IEEE（Institute of Electrical and Electronics Engineering） 电子电器工程师协会
IETF（Internet Engineering Task Force） 因特网工程任务组
IIC（Interface Integrated Circuit） 集成接口电路
IIS（Internet Information Service） 互联网信息服务
IM（Instant Message） 即时消息
IOCP（Input/Output Control Program） 输入/输出控制程序
IP（Internet Protocol） 因特网协议
IP（Intellectual Property） 知识产权
IPC（Inter-Process Communication） 进程间通信
IPSE（Integrated Project Support Environments） 集成工程支持环境
IPX（Internet Packet eXchange） 网间分组交换

IRC（Internet Relay Chat） 在线聊天系统
ISDN（Integrated Service Digital Network） 综合业务数字网
ISO（International Standard Organization） 国际标准化组织
ISP（Internet Service Provider） 因特网服务提供者
IT（Information Technology） 信息技术
ITU（International Telecom Union） 国际电信联盟

J

JAR（Jump Address Register） 转移地址寄存器
JF（Journal File） 日志文件
JIVA（Joint Intelligence Virtual Architecture） 联合智能虚拟体系结构
JOF（Job Output File） 作业输出文件
JDBC（Java Database Connectivity） Java 数据库互连
JDK（Java Developer's Kit） Java 开发工具包
JPEG（Joint Photographic Experts Group） 联合图像专家组
JSP（Java Server Page） Java 服务器页面技术

K

KA（Knowledge Acquistion） 知识采集
KB（KiloByte） 千字节
KBC（Keyboard Controller） 键盘控制器
KBMS（Knowledge Base Management System） 知识数据库管理系统
KDL（Knowledge Definition Language） 知识定义语言
KIU（Keyboard Interface Unit） 键盘接口单元
KM（Knowledge Module） 知识模块
KNS（Key Notarization System） 密钥公证系统
KWIC（Key Word In Context） 文本关键字索引法
KBPS（KiloBits Per Second） 每秒千比特
KMS（Knowledge Management System） 知识管理系统
KQML（Knowledge Query and Manipulation Language） 知识查询和管理语言

L

LADS（Local Area Distributed System） 本地分布式系统
LADT（Local Area Data Transport） 本地数据传输
LAN（Local Area Network） 局域网
LARS（Learning And Recognition System） 学习和识别系统
LAT（Local Area Transport） 本地传输
LAWN（Local And Wireless Network） 局域无线网络
LC（Language Converter） 语言转换器
LCD（Liquid Crystal Display） 液晶显示器
LCS（Loop Control Statement） 循环控制语句
LED（Light Emitting Diode） 发光二极管

LLC (Logical Link Control sub-layer) 逻辑链路控制子层
LMB (Left Mouse Button) 鼠标器左按钮
LN (Loop Network) 环型网络
LP (Linear Programming) 线性规划
LPS (Lines Per Second) 每秒行数
LR (Linear Relationship) 线性关系
LSIC (Large Scale Integration Circuit) 大规模集成电路

M

MA (Multiple Access) 多路访问
MADE (Multimedia Application Development Environment) 多媒体应用开发环境
MAN (Metropolitan Area Network) 城域网
MB (Megabytes) 兆字节（存储容量单位）
MC (Memory Card) 存储卡片
MCA (Micro Channel Architecture) 微通道结构
MDA (Monochrome Display Adaptor) 单色显示适配器
MFM (Modified Frequency Modulation) 改进调频制
MHz (Megahertz) 兆赫（频率单位）
MIB (Management Information Bass) 管理信息库
MIDI (Music Instrument Digital Interface) 音乐设备数字接口
MIMD (Multiple Instruction Stream, Multiple Data Stream) 多指令流，多数据流
MIPS (Million Instruction Per Second) 每秒百万条指令
MIS (Management Information System) 管理信息系统
MIRS (Multimedia Information Retrieval System) 多媒体信息检索系统
MISD (Multiple Instruction Stream, Single Data Stream) 多指令流，单数据流
MMC (Microsoft Management Console) 微软管理控制台
MMI (Man Machine Interface) 人机界面
MMU (Memory Management Unit) 内存管理单元
MP3 (MPEG Audio Layer3) MP3 音乐
MPC (Multimedia PC) 多媒体计算机
MTBF (Mean Time Between Failure) 平均故障间隔时间
MTV (Musical Television) 音乐电视
MUD (Multiple User Dimension) 多用户空间
MVC (Multimedia Video Card) 多媒体视频卡
MW (Multi-Window) 多窗口
MXL (Multiplex Link) 多路复用链路

N

NA (Network Adapter) 网络适配器
NAF (Network Access Facility) 网络接入设备
NAP (Network Access Point) 网络接入点
NBBS (Network Board Band Services) 网络宽带业务

NC（Network Computer） 网络计算机
NCU（Network Control Unit） 网络控制单元
NCSC（National Computer Security Center） 国家计算机安全中心
NDAS（Net Dynamics Application Server） 网络动态应用服务器
NDIS（Network Device Interface Specification） 网络设备接口规范
NFS（Network File System） 网络文件系统
NIS（Network Information Services） 网络信息服务
NIST（National Institute of Standards and Technology） 国家标准化与技术研究所
NLOS（Natural Language Operating System） 自然语言操作系统
NORMA（No Remote Memory Access（multip- rocessor）） 非远程内存访问（多处理器）
NRU（Not Recently Used） 非最近使用
NSP（Name Server Protocol） 名字服务器协议
NUI（Network User Identifier） 网络用户标识符

O

OA（Office Automation） 办公自动化
OAI（Open Application Interface） 开放应用程序接口
OBI（Open Business on the Internet） 互联网上的开放商行
OCR（Optical Character Recognition） 光学字符识别
ODBC（Open Database Connectivity） 开放式数据库互连
ODI（Open Data-link Interface） 开放式数据链路接口
ODL（Object Data Language） 对象数据语言
OEM（Original Equipment Manufactures） 原始设备制造厂家
OH（Operator Handbook） 操作员手册
OLE（Object Linking and Embedding） 对象链接与嵌入
OMG（Object Management Group） 对象管理组织
OMR（Optical Mark Recognition） 光学符号识别技术
OODB（Object-Oriented Database） 面向对象的数据库
OOP（Object Oriented Programming） 面向对象程序设计
ORG（Object Request Broker） 对象请求代理
OS（Operating System） 操作系统
OSI（Open System Interconnect Reference Model） 开放式系统互连参考模型
OSPF（Open Shortest Path First） 开发最短路径优先
OWL（Object Window Library） 对象视窗库
OWS（Online Workgroup Server） 在线工作组服务器
OWT（One-Way Transmission） 单向传输

P

PAB（Personal Address Book） 个人地址簿
PANS（Private Access Network System） 专用接入网系统
PBS（Public Broadcasting Service） 公共广播服务
PC（Personal Computer） 个人计算机

PCI（Peripheral Component Interconnect） 外部部件互连
PCN（Personal Communication Network） 个人通信网
PDA（Personal Digital Assistant） 个人数字助理
PDF（Portable Document Format） 便携式文档格式
PDN（Public Data Network） 公共数据网
PKCS（Public Key Cryptography Standard） 公共密钥密码标准
PKI（Public Key Infrastructure） 公共密钥基础结构
PMMU（Paged Memory Management Unit） 页面存储管理单元
PnP（Plug and Play） 即插即用
POP（Post Office Protocol） 邮局协议
POS（Point Of Sale） 销售点，自动收款机
POST（Power-On Self-Test） 加电自检
PPP（Point to Point Protocol） 点到点协议
PPSN（Public Packed-Switched Network） 公用分组交换网
PROM（PRogrammable ROM） 可编程只读存储器
PSDN（Public Switched Data Network） 公共交换数据网
PSDN（Packet Service Digital Network） 分组业务数字网
PWT（Personal Wireless Telecommunications） 个人无线通信

Q

QAM（Queue Access Method） 排队存取法
QAS（Quick Access Method） 快速存取存储器
QC（Quality Control） 质量控制
QI（Quality Index） 质量指标
QL（Query Language） 查询语言
QLP（Query Language Processor） 查询语言处理器
QoS（Quality of Service） 服务质量
QR（Quick Response） 快速响应
QTR（Quality Technical Report） 质量技术报告
QUIP（Quad-In-Line Package） 四列直插封装

R

RA（Random Access） 随机存取
RACS（Remote Automatic Control System） 远程自动控制系统
RAI（Remote Application Interface） 远程应用程序界面
RAM（Random Access Memory） 随机存储器
RAM（Real Address Mode） 实地址模式
RAID（Redundant Arrays of Inexpensive Disks） 冗余磁盘阵列技术
RAW（Read After Write） 写后读
RCP（Remote CoPy） 远程复制
RDA（Remote Data Access） 远程数据访问
RF（Radio Frequency） 无线射频

RFQ (Request for Quotation) 请求引证
RFS (Remote File Service) 远程文件服务
RGB (Red, Green, Blue) 三原色（红色、绿色、蓝色）
RIP (Raster Image Processor) 光栅图像处理器
RISC (Reduced Instruction Set Computer) 精简指令集计算机
RMB (Right Mouse Button) 鼠标器右键
ROM (Read Only Memory) 只读存储器
RPC (Remote Procedure Call) 远程过程调用
RVT (Routing Vector Table) 路由矢量表
RX (Remote Exchange) 远程交换机

S

SACE (System Auxiliary Control Element) 系统辅助控制单元
SAF (Secure Access Firewall) 安全接入防火墙
SAN (Small Area Network) 小区网络
SAP (Service Access Point) 服务访问点
SAS (Self-Adaptive System) 自适应系统
SCSI (Small Computer System Interface) 小型计算机系统接口
SDD (System Description Document) 系统描述文档
SDLC (Synchronous Data Link Control) 同步数据链路控制
SDK (Software Development Kit) 软件开发工具箱
SDL (Specification and Description Language) 技术规范和描述语言
SES (Signal Encrypt System) 信号加密系统
SFT (System Fault Tolerance) 系统容错
SGML (Standard Generalized Markup Language) 标准通用标记语言
SHTTP (Secure Hype Text Transfer Protocol) 安全超文本传递协议
SIMD (Single Instruction Stream, Multiple Data Stream) 单指令流，多数据流
SIMM (Single Inline Memory Module) 单列直插式内存模块
SIO (Serial Input / Output) 串行输入/输出
SISD (Single Instruction Stream, Single Data Stream) 单指令流，单数据流
SMB (Server Message Block) 服务器消息块
SMDS (Switch Multi-megabit Data Services) 交换多兆位数据服务
SMPC (Shared Memory Parallel Computer) 共享存储器并行计算机
SMTP (Simple Mail Transport Protocol) 简单邮件传输协议
SNA (System Network Architecture) 系统网络结构
SNMP (Simple Network Management Protocol) 简单网络管理协议
SNTP (Simple Network Time Protocol) 简单网络时间协议
SOAP (Simple Object Access Protocol) 简单对象访问协议
SOHO (Small Office / Home Office) 小型办公室和家庭办公室
SONET (Synchronous Optic Network) 同步光纤网
SPA (Secure Password Authentication) 安全口令认证

SPC(Stored-Program Control) 存储程序控制
SQL(Structured Query Language) 结构化查询语言
SSIC(Small Scale Integration Circuit) 小规模集成电路
STA(Spanning Tree Algorithm) 生成树算法
STB(Set Top Box) 机顶盒，顶置盒
STDM(Synchronous Time Division Multiplexing) 同步时分复用
SUS(Speech Understanding System) 语音理解系统

T

TCB(Transmission Control Block) 传输控制块
TCM(Terminal to Computer Multiplexor) 终端到计算机的多路转接器
TCP(Transmission Control Protocol) 传输控制协议
TCP/IP(Transmission Control Protocol/Internet Protocol) 传输控制协议/网间协议
TDM(Time Division Multiplexing) 时分多路复用
TDMA(Time Division Multiplexing Address) 时分多址技术
TIFF(Tagged Image File Format) 有标签的图形文件格式
TIG(Task Interaction Graph) 任务交互图
TI-RPC(Transport-Independent Remote Procedure Call) 独立传输远程过程调用
TLI(Transport Layer Interface) 传输层接口
TMN(Telecommunication Management Network) 电信管理网
TNC(Terminal Network Controller) 终端网络控制器
TSC(Time-Shared Computer) 分时计算机
TSR(Terminate and Stay Resident) 终止并驻留
TSS(Telecommunication Switching System) 电信交换系统
TTL(Transistor-Transistor Logic) 晶体管-晶体管逻辑电路
TWC(Two-Way Communication) 双向通信
TWX(TeletypeWriter Exchange) 电传电报交换机

U

UART(Universal Asynchronous Receiver Transmitter) 通用异步收发器
UBCS(Unified Byte Code System) 统一字节编码系统
UDF(Universal Disk Format) 通用磁盘格式
UDI(Uniform Driver Interface) 统一驱动器接口
UDLC(Universal Data Link Controller) 通用数据链路控制器
UDP(User Datagram Protocol) 用户数据报协议
UES(User's Electronic Signature) 用户电子签名
UIMS(User Interface Management System) 用户接口管理程序
UNI(User Network Interface) 用户网络接口
UPS(Uninterruptible Power Supply) 不间断电源
URI(Uniform Resource Identifier) 环球资源标识符
URL(Uniform Resource Locator) 统一资源定位器
USB(Universal Serial Bus) 通用串行总线

UTP（Unshielded Twisted-Pair） 非屏蔽双绞线
UWA（User Work Area） 用户工作区

V

VAN（Value Added Network） 增值网络
VAP（Value-Added Process） 增值处理
VAS（Value-Added Server） 增值服务
VAX（Virtual Address eXtension） 虚拟地址扩充
VCPI（Virtual Control Program Interface） 虚拟控制程序接口
VDD（Virtual Device Drivers） 虚拟设备驱动程序
VDI（Video Device Interface） 视频设备接口
VDT（Video Display Terminals） 视频显示终端
VDU（Visual Display Unit） 视频显示单元
VGA（Video Graphics Adapter） 视频图形适配器
VIS（Video Information System） 视频信息系统
VLAN（Virtual LAN） 虚拟局域网
VLSI（Very Large Scale Integration） 超大规模集成
VMM（Virtual Memory Management） 虚拟内存管理
VMS（Virtual Memory System） 虚拟存储系统
VOD（Video On Demand） 视频点播系统
VOS（Virtual Operating System） 虚拟操作系统
VPN（Virtual Private Network） 虚拟专用网
VR（Virtual Reality） 虚拟现实

W

WAN（Wide Area Network） 广域网
WAE（Wireless Application Environment） 无线应用环境
WAP（Wireless Application Protocol） 无线应用协议
WB（Warm Boot） 热启动
WDM（Wavelength Division Multiplexing） 波分多路复用
WDP（Wireless Datagram Protocol） 无线数据包协议
WFW（Windows for Workgroups） 工作组窗口
WML（Wireless Markup Language） 无线标记语言
WORM（Write Once, Read Many time） 写一次读多次光盘
WPS（Word Processing System） 文字处理系统
WSDL（Web Services Description Language） 服务描述语言
WSP（Wireless Session Protocol） 无线会话层协议
WWAN（Wireless Wide Area Network） 无线广域网
WWVC（World Wide Video Communication） 互联网视频通信
WWW（World Wide Web） 万维网

X

XDR（eXternal Data Representation） 外部数据表示

XGA（eXtended Graphics Array） 扩展图形阵列
XIOS（eXtensible Input/Output System） 扩展的输入/输出系统
XML（eXtensible Markup Language） 可扩展标记语言
XMM（eXtensible Memory Manager） 扩展内存管理器
XMS（eXtended Memory Specification） 扩展存储器规范
XQL（eXtensible Query Language） 可扩展查询语言

Z

ZA（Zero and Add） 清零与加指令
ZAC（Zero Address Code） 零地址码
ZAI（Zero Address Instruction） 零指令地址
ZBR（Zone Bit Recording） 零位记录制
ZC（Zero Compression） 零压缩
ZIP（Zone Information Protocol） 区域信息协议
ZTAS（Zero Wait State Computer） 瞬时自动化系统

参考文献

[1] 吕云翔. 计算机英语实用教程. 北京：清华大学出版社，2015.
[2] 王小刚. 计算机专业英语. 4版. 北京：机械工业出版社，2015.
[3] 金志权，张幸儿. 计算机专业英语教程. 6版. 北京：电子工业出版社，2015.
[4] 张芳芳. 计算机网络专业英语. 北京：电子工业出版社，2014.
[5] 苏兵，张淑荣. 计算机英语. 2版. 北京：化学工业出版社，2014.
[6] 卜艳萍，周伟. 计算机专业英语. 2版. 北京：人民邮电出版社，2012.
[7] 赵桂钦. 电子与通信工程专业英语. 北京：清华大学出版社，2012.
[8] 卜艳萍，周伟. 计算机专业英语. 北京：清华大学出版社，2010.
[9] 甘艳平，等. 信息技术专业英语. 北京：清华大学出版社，2009.
[10] 刘兆毓. 计算机英语. 4版. 北京：清华大学出版社，2009.
[11] 丛书编委会. 计算机类专业英语. 北京：中国电力出版社，2008.
[12] 张勇. 计算机专业英语. 北京：北京大学出版社，2008.
[13] 邱仲潘，等. 计算机英语. 北京：中国铁道出版社，2006.
[14] 武马群. 计算机专业英语. 北京：北京工业大学出版社，2005.
[15] 张玲等. 计算机专业英语. 北京：机械工业出版社，2005.